PENGUIN R

THE PENGUI
TELECOM

John Graham has extensive experience in the computer
and telecommunication industries. He has held a number
of key appointments in ICL from 1964, including that of
Manager, Corporate Systems, and Regional Manager
for Dataskil, then a software house subsidiary of ICL.
Later he became Systems Engineering Manager, responsible for all technical support activities in ICL's International Division, and finally Managing Director of
ICL's marketing subsidiary in the Far East. In 1979
he joined Aregon International Limited, where as Marketing Director he was responsible for the development
and implementation of its international business strategy
in the videotex field. He is currently Managing Director
of Aregon Products. He is the author of *Systems Analysis
in Business, Making Computers Pay* and *The Penguin
Book of Personal Computing* and co-author (with
Anthony Chandor and Robin Williamson) of *The
Penguin Dictionary of Computers*.

THE PENGUIN DICTIONARY OF

TELECOMMUNICATIONS

JOHN GRAHAM

PENGUIN BOOKS

Penguin Books Ltd, Harmondsworth, Middlesex, England
Viking Penguin Inc., 40 West 23rd Street, New York, New York 10010, U.S.A.
Penguin Books Australia Ltd, Ringwood, Victoria, Australia
Penguin Books Canada Ltd, 2801 John Street, Markham, Ontario, Canada L3R 1B4
Penguin Books (N.Z.) Ltd, 182–190 Wairau Road, Auckland 10, New Zealand

First published 1983
Published simultaneously by Allen Lane
Reprinted 1985

Made and printed in Great Britain by
Richard Clay (The Chaucer Press) Ltd,
Bungay, Suffolk
Filmset in Monophoto Times by
Northumberland Press Ltd, Gateshead

CONTENTS

PREFACE

Our world has become an increasingly complex place in which, as individuals, we are very dependent on other people and upon organizations. An event in some distant part of the globe can rapidly and significantly affect the quality of life in our home country.

This increasing interdependence, on both a national and international scale, has led us to create systems which can respond immediately to dangers, enabling appropriate defensive or offensive actions to be taken. These systems are operating all around us in military, civil, commercial and industrial fields.

The electronic computer is at the heart of many such systems, but the role of telecommunications is no less important. As we proceed through the 1980s, there will be a further convergence between the technologies of computing and telecommunications. The changes will be dramatic:

> the paperless office
> the database society
> the cashless society
> the office at home

We cannot doubt that the economic and social impact of these concepts will be very significant. Already, advanced systems of communication are affecting both the layman and the technician. Complex functions are being performed by people using advanced terminals which are intended to be as easy to use as the conventional telephone.

The aim of the book

Telecommunications principles are becoming increasingly important in education at undergraduate and graduate levels. All those engaged in the fast-growing systems industries are finding that a knowledge of telecommunications is essential to solve today's problems; this knowledge is also vital to those who are responsible for managing distributed organizations.

It is the aim of this book to help people with the terminology of this subject and, at the same time, to provide a convenient source of reference to basic tele-communication principles.

In common with many other modern technologies, the subject of telecommunications has developed a language of its own. At first sight, the words appear to be intelligible English but, of course, with the inevitable sprinkling of initials such as FSK (*frequency shift keying*). The close association with the computer industry gives rise to some major new fields of technical complexity, e.g. *packet switching*.

This book sets out to define such terms and, as far as possible, to adhere to definitions in keeping with international practice. The scope of the publication is wide, dealing with fundamental concepts in telephone and telegraph communications, switched communications systems, broadcast systems, and narrow and broad bandwidth systems.

The intention is to make the material intelligible to the layman as well as interesting and informative for the technician. It is hoped that it will assist readers to deal with other technical documents that they have to read, and prove useful for browsing.

With these objectives in mind, I have aimed at a style which is informative rather than precise. There are also many cross-references to other definitions which complement, or contrast with, the definition of a particular term. In many definitions, it has been necessary to use other technical terms, which are themselves defined elsewhere in the book. Such terms have been printed in italics as they arise in the text.

Spellings and sequence

Certain words in this book have been spelt in accordance with the conventions adopted in the industry which owes so much to North American influences: e.g. computer program, analog computer, toll center. I have also preferred to use multiplexor rather than multiplexer.

The book has been arranged according to strict alphabetic sequence of each word and, where a term is composed of two or more words, the sequence is determined by the first word, and then by the second, and so on: e.g.

> call accepted signal
> call request signal
> call set-up
> called line

In this way I have tried to spare the reader the task of interpreting the sequence in which a computer sorts strings of text and symbols. A few terms begin with a numeral, and I have then given the numeral the exact significance of the corresponding word; thus the term *2-bit error* appears immediately before *two condition code*.

Many people and organizations have provided help in the preparation of the material, and I particularly wish to mention Jeremy Wood, who helped with the illustrations, and Tracey Mitchell, who typed some of the early drafts. Above all I must thank my wife, Dianne, who gave so much of her time, energy, patience, and good humour to read, correct, index, and sequence all the text. She took care of all cross-references and made the preparation of the words such a straightforward operation. She also typed the final manuscript.

John Graham
Worplesdon
October 1981

NOTE

See references are denoted by an arrow: ⇨
See also references are denoted by a double arrow: ⇨⇨

A

abbreviated address calling A system in which a subset of *address* characters is required to establish a *call* from a *terminal*, and in which the *network* expands the abbreviated address to the full address. ⇨ *abbreviated dialling service*.

abbreviated dialling service A service made available to *users* of a switched *network*, in which non-numerical *codes* are available to represent groups of *digits* and specify code *prefixes*. This facility enables a caller to use an *address* having fewer *characters* than the full address of the *called subscriber*.

abbreviated number An *address* sent by a caller to indicate the number of a *subscriber* to be called but which is preceded by a non-numerical *prefix* indicating the use of an *abbreviated dialling service*.

abbreviated prefix dialling In the operation of a telephone service, the use of a non-numerical code to indicate that the following *digits* are an *abbreviated number*.

AC ⇨ *alternating current*.

AC signal A *signal* in which the direction of the current is reversed in accordance with a frequency; e.g. a *speech signal* varies within a range of *audio frequencies* according to vibrations generated by a human voice.

accentuated contrast A technique used in *facsimile* to transmit information relating to documents, in which parts of a picture having a *luminance* less than a specified value are transmitted as black, and parts having a greater luminance as white.

access-barred signal A *signal* sent back to a *calling terminal*, to indicate that the caller is not allowed to be connected to the *called*

location; e.g. he is not a member of a *closed user group* having access to that location.

access code A *prefix* to an *address* which enables a caller to obtain a connection to a specific service; e.g. to the international automatic *telex* network.

access path A path providing communication between two *terminals*. In a *telephone system*, or a *data network*, there may be a number of possible paths between two terminals, but for a particular *call* a specific set of physical transmission resources must be allocated to provide a specific access path. Sometimes the path is established via one or more intermediate *nodes* in the *network*.

access path control Usually in a communications system there is more than one set of physical resources available to provide an *access path* from a *calling location* to a *called destination*. There may also be a choice of intermediate *nodes* through which the *call* can be *routed*. Access path control procedures are concerned with the selection of a particular *transmission path*. The procedures usually reside in logic built into the *exchanges* or nodes of the network, and the paths are selected automatically according to conditions prevailing.

access point A physical point of connection to a *network* which defines the limits of a *line*. For example, only one *line access point* can exist at each end of an international line.

accounting rate A term which relates to communications *traffic* between zones controlled by different telecommunications authorities, and which is used for the establishment of international accounts. The rate is expressed as a charge per *traffic unit*; e.g. per *word*, per minute, depending upon the type of service.

accounting rate quota That part of the accounting rate given to a *telecommunications* authority for *traffic* between it and another such authority, and relating to the facilities utilized in each country. This quota is fixed by predetermined agreement between the authorities concerned.

accounting revenue division procedure A procedure whereby revenue relating to a communications service is shared between the *telecommunications* authorities providing the service at each end.

ACK ⇨ *acknowledge.*

acknowledge (ACK) An international *transmission control code* which is returned by a *receiving terminal* to a *transmitting terminal* to acknowledge that a *frame* of information has been correctly received. Contrast with *negative acknowledge (NAK).*

acknowledgement indicator A *signal* which is used to indicate whether or not an error has been detected in a *frame*, or some other unit of information. The signal is transferred back from the *receiving* to the *transmitting terminal.* ⇨ *ACK* and *NAK.*

acknowledgement signal unit A part of a *block* of signal information transmitted over a *public data network* and containing information to indicate whether the other *signal* units in the block were correctly received.

acoustic coupler A *transducer* for coupling a *data terminal (DTE)* to a telephone *network* in which *data signals* from the *terminal* are converted to *sound waves*, and sounds received from the telephone *line* and intended for the terminal are converted to data signals. The DTE can thus communicate using a telephone *handset* with a distant terminal or *host processor.*

acoustically coupled modem A special *modem* which enables tones generated by a *data terminal* to be transferred to a tele-

phone *handset*, thereby providing *data communication* over the *public switched telephone network (PSTN).*

active lines In forming a TV picture, an *electron beam* traces a pattern of *lines* known as a *raster.* Some lines do not contain *luminance* information, (e.g. *field flyback*). The lines which do contain luminance information are known as active lines.

actual final route The specific path followed by a *call* over an international *switched* telephone *network*, to be contrasted with the *theoretical final route.* The actual final route may, or may not, use all or part of the theoretical final route.

ACU Abbreviation of *acknowledgement signal unit.*

adaptive channel allocation A technique used in *intelligent time division multiplexing* over *public data networks* in which the information capacity of a *channel* is determined in response to demand, rather than being predetermined. Same as *dynamic channel allocation.*

adaptive equalizer A device which counteracts *delay distortion* affecting *analog signals* in transmission circuits, and constructed to respond appropriately at different speeds and operating conditions. ⇨ *equalization.*

adaptor A device used to *interface* a *data terminal (DTE)* to a *channel*, and to carry out *code* and *data rate* conversion to enable the terminal to be compatible with the communication standards required.

ADCP Abbreviation of *advanced data communication protocol.*

address That part of a *signal* which defines the destination for a *call.*

address-complete signal In a *switched network*, a *signal* sent in a *backward channel*, indicating that all *digits* of an *address* have been received and that the *call* should be charged on answer by the *called party*. Compare with *address-incomplete signal*.

address digits Elements of signal information representing *digits* 1, 2 to 9 or 0 which together define the telephone number of a *called party*, or a *calling party*.

address field A *field* of information forming part of a *frame* which indicates a unique *address* for a *station* connected to a *network*. ⇨ *destination address field* and *source address field*.

address-incomplete signal In a *switched network*, a *signal* sent by an *exchange* in a *backward channel* indicating that the number of *address digits* received is not sufficient to set up a *call*.

address information Part of the information contained in a *message* or *packet*, and intended to provide to the *control* system of the *network* the necessary information to identify the *user* to whom the message or packet is to be sent, and to allow a *route* to be selected.

address message A *message* sent in a *forward direction* over a *public data network* and containing the information necessary to *route* and connect a *call* to a *called party*.

address signal A *signal* representing just one *character* element of the complete *address*, (e.g. 9 or 5). For each address, a succession of *address signals* will occur; also, depending upon the type of *network*, an *end-of-address signal*.

addressing Most communication systems use the concept of several *users* sharing the same *transmission lines* (e.g. by *circuit switching*, or by *time switching* in a *time division multiplexing* environment). Some method is required to make connections between specific users, or to transmit *packets* of information between specific users. To achieve connections, addressing information must be generated as part of

the *call establishment* process, or be included in the *data* transmitted. Using this addressing information, a *routing* operation takes place to connect the required *terminals*, and the path remains available until the particular transmission has been completed.

addressing flexibility A quality desired in the construction of *networks* to allow *frames* of information to be directed to a specific *user*, a group of users, or to all users on the network.

advanced data communication protocol (ADCP) A standard developed by the American National Standards Institute, which specifies *error control* protocols for *data transmission*, to ensure that information received at a *terminal* is a replica of the *bit-pattern* transmitted. ⇨ *high level data link control (HDLC)*.

aerial A part of a radio system which either radiates *signals* into the atmosphere or receives *radio signals* from the atmosphere. ⇨ *antenna*.

AGC ⇨ *automatic gain control*.

alphabet Relating to a *data communications code* and defining an agreed set of *characters* or symbols and the *signals* which represent them. ⇨, for example, *International Alphabet No. 5*.

alphabetic telegraphy A conventional form of *telegraphy* using an *alphabetic telegraph code* (e.g. *International Telegraph Alphabet No. 2*) transmitted between two *terminals*. Such codes are used to represent letters, figures and punctuation marks, each as a unique pattern of pulses, and are known as *data communication codes*. An example is shown in Appendix 1. The term 'alphabetic telegraphy' is used to distinguish it from other forms of telegraphy; e.g. using *facsimile* methods.

alphageometric coding A system of coding for storing, transmitting and displaying *graphic information*; e.g. in a *videotex system* or *computer aided design system*. Such a system can be defined to enable very complex colour graphic images to be displayed. The description given below relates to one of the early videotex systems to offer *high resolution graphics*; it is a typical example of alphageometric coding.

The system displays *text* and *graphics* on a *television receiver* using a 525-line system and offering a format of 320 × 240 *picture elements*. Because the system is used to transmit images over a *narrow bandwidth* offered by the *public switched telephone network*, some method is required to compress the picture information into *codes* which can be transmitted efficiently. These codes, known as *picture description instructions* (*P D Is*), are defined as primitive components such as an arc, a line, a point, or a polygon. All pictures are made up of combinations of these primitives and are created by the powerful logic incorporated in the *terminal* acting upon PDIs it receives from the *telephone channel*.

The technique offers a high degree of flexibility, and complex diagrams can be compressed to reduce transmission time. Such systems produce much finer image definition than *alphamosaic coding* but take longer to transmit on a telephone channel and require more expensive *decoders* in the terminal. As the cost of micro-electronic technology improves, it seems likely that alphageometric coding systems will gain in popularity for public services. Compare also with *alphaphotographic display system*.

alphamosaic coding A system of coding used in *videotex systems* to enable *text* and *graphic information* to be stored in a central *computer*, and transmitted and displayed on a *videotex terminal*. The graphic displays produced by this technique are of relatively *low resolution* but it provides an extremely cheap and effective way of conveying graphics in such a public service system. The pictures are formed of individual *graphic characters*; e.g. up to 63 different characters are defined from a character matrix of 2 × 3 elements. This is the most common method of coding graphics in *public videotex systems* which are being introduced throughout Europe in the 1980s. Compare with *alphageometric coding* and *alphaphotographic display system*.

alphamosaic graphics Graphic images composed of *graphic characters* combined to form a picture in a *videotex system*. ⇨ *alphamosaic coding*.

alphanumeric Any set of symbols consisting of alphabetic *characters* and special symbols as well as *numbers*.

alphaphotographic display system A graphic display system providing *high resolution* images comparable in quality to a colour television picture. An example is described under the entry *Picture Prestel*. For comparison, ⇨ *alphamosaic coding* and *alphageometric coding*.

alternate digit inversion A technique used to prevent loss of *synchronization* in *repeaters* used for regenerating *digital signals*. In order to identify incoming pulses, some repeaters have internal *clocks* which are synchronized to the incoming stream of *bits*. During periods of low activity, the incoming stream will be all zeros and there is a tendency under these conditions for the clock to lose synchronization. Deliberate inversion of the *time slots* relating to alternate bits is used to prevent the occurrence of continuous zeros and thus maintain synchronization.

alternate mark inversion (A M I) A system for line transmission in which the *mark* condition of a coded *signal* is represented alternately by a positive and negative voltage of equal *amplitude*, and in which *space* is represented by zero amplitude.

alternate mark inversion violation An error condition in which two consecutive *mark* signals have the same polarity.

alternate path routing In a *network*, a *routing* involving the selection of an *access* *path* other than the basic or normal theoretical access path. Usually selected because of failure in the normal access path, or because an intervening *node* or *transmission link* is heavily overloaded.

alternating code A *code* in which *bits* or *characters* may be represented under a given set of rules by two different conditions; e.g. '1' *digit* alternately represented as a positive or negative pulse. This technique assists maintenance of timing in a long sequence of digits. An example is given under *alternate mark inversion*. Also known as *paired-disparity code*.

alternating current (AC) An electric circuit in which the direction of the current is reversed. For example, in Britain the mains electrical power supply varies at a frequency of 50 cycles per second. Other examples of simple waveforms are shown in Appendix 3 and Appendix 7. *Electrical signals* to represent sounds such as speech and music are *AC signals*, but they have complex forms consisting of many harmonic frequencies and varying *amplitudes*.

alternative coding ⇨ *alternating code*.

alternative routing The optimum, or *basic routing*, for a *call* in a *switched system* is one which uses least *trunk circuits*.
  A *routing*, other than the basic routing for a call, is known as an alternative routing. ⇨ *exchange hierarchy*.

alternative routing indicator Information sent in the *forward direction* of a *call* over a *public data network* indicating that an *alternative routing* has been used.

AM Abbreviation of *amplitude modulation*.

AMI ⇨ *alternative mark inversion*.

amount of traffic carried In a *switched system* the amount of *traffic* carried is the sum of the *holding time* of all the *calls* made in any given period over a given *trunk*

or through a given *exchange*. Usually measured in hours.

amplification The process involved in increasing the power of a *signal*. This is often done to overcome loss of power which arises due to the electrical resistance in the materials used to construct a *transmission channel*.

amplifier A device used to increase the power of an *electrical signal* to overcome cable losses and to deliver a signal of the appropriate *minimum signal level* to a device at the *message* destination.

amplifier bandwidth The effective *frequency band* over which an *amplifier* operates; e.g. in a telephone *circuit*, amplifiers are required to amplify *message signals* derived from *voice signals* spoken into a *microphone*. For this purpose, a *bandwidth* of from 0·3 to 3·4 kHz is suitable. Very often amplifier bandwidth is selected to filter out *noise* occurring outside the critical band occupied by the message signal.

amplifier gain The factor by which an *amplifier* increases the power of a *signal* – a ratio of the input to output power. ⇨ *gain*.

amplitude A measure of the strength or loudness of a *signal*. Two *electrical signals* may have the same frequency but, if one is of greater amplitude than the other, it will be represented by higher voltage peaks than the other. Amplitude is the peak value (in a positive or negative direction) of an *alternating current* signal.

amplitude modulation (AM) A form of modulation in which the *amplitude* of a *carrier wave* is made to vary in sympathy with the frequency of a *signal* which is to be transmitted. Any form of modulation in which different conditions are indicated by currents of different amplitude. ⇨ *modulation* and Appendix 7.

analog A direct representation of a phenomenon in another form; e.g. the representation of voice-sounds as electrical *audio signals.* ⇨ *analog electrical signal.*

analog data channel A *channel* designed to carry *analog electrical signals* as distinct from *digital signals.* Although there is an increasing tendency to carry *data traffic* on high speed digital circuits, the use of telephone *circuits* to carry *data* represented in analog form is still common – a *modem* is used to convert digital signals to and from the analog format expected on the *line.*

analog electrical signal An *electrical signal* which directly represents another form of energy or activity. The most obvious example is the representation of *sound waves* in electrical form as an *audio signal* which directly corresponds in frequency to the sound waves that it represents.

analog line terminating system A unit which converts *analog signals* representing speech and other forms into digital form and vice versa. ⇨ *pulse code modulation.*

analog repeater An *amplifier* designed to regenerate *analog signals* and used at regular intervals along a *transmission path* to overcome the problem of *attenuation.* Compare with *digital repeater.*

analog signal ⇨ *analog electrical signal.*

analog techniques The use of one medium to directly represent a phenomenon or activity occurring in another medium; e.g. the use of *electromagnetic waves* to represent *sound waves.*

ancillary device A device (e.g. a *hard copy printer*) which is attached to, and under the control of, another device such as a *terminal* or *node.*

AND gate A *circuit* with two (or more) inputs and a single output in which an output signal is provided if all inputs are present.

anisochronous system A communication system in which a common *time interval* is not established between *transmitting stations* and *receiving stations.* The timing is established by *start* and *stop bits* which *frame* each transmitted unit of *data*; e.g. *character.* Strictly speaking, a system in which there is an integral number of time units within a given unit of data (e.g. character) but not necessarily between successive units in the same *message.* More commonly referred to as an *asynchronous system.* Contrast with *isochronous system.*

annual charge ratio A ratio used to express the quality of service and charges in the operation of a *telephone system.* The ratio is determined by taking the annual charge of one additional *circuit* on the *alternative route*, and dividing this sum by the annual charge of one additional circuit on the *high-usage route.*

answer back simulator A device, or a program routine, which is not part of a *teleprinter* but which behaves in the same way as an *answer back unit* in response to a specific *who are you* signal.

answer back unit A device in a *teleprinter* which automatically sends an identification signal to a *calling terminal* in response to a specific *who are you* signal.

answer signal A *signal* sent in the *backward direction* when a *call* is answered. The signal usually initiates the next stage in the progress of the call. In public service systems this signal also triggers the metering process which is used to calculate the charge for the call.

answer signal, charge A *signal* sent in the *backward direction* when a *call* is answered, and which is used to start metering the call to calculate the charge to be made to the *calling party.*

answer signal, no charge A *signal* sent in the *backward direction* when a *call* is answered, and which is used to indicate that this call is not subject to any charge;

e.g. a call made to public emergency services.

answering, automatic ⇨ *automatic answering*.

answering, manual ⇨ *manual answering*.

answering time The time interval between the completion of the transmission of a *calling signal* and the answer by an operator or by an *automatic answering* system at the distant location.

antenna A device constructed to transmit or receive *radio waves*. The physical characteristics of the antenna determine its suitability to send or receive radio waves in a particular *frequency band*. Antenna also may be *directional* or *non-directional*.

A directional antenna used for transmission radiates a greater degree of energy in a particular direction, and is usually arranged to beam radio waves in the direction of a particular *receiving terminal*. Thus, it is used in point-to-point communication.

application The task to which a system is applied and the procedures which relate to the execution of it. For example, a *computer* system and a *telecommunications* system may be used to handle an airline ticket reservation system. This is an example of an application.

application code The programs which perform a particular data processing task for *users*, as distinct from the *operating system* or other general-purpose *software* forming part of the *computer* environment.

application layer Relating to the *architecture* of a communications system and defining the levels of *hardware* or *software* which relate to the *application* to be performed rather than the levels which are concerned with managing and controlling *data transmission*. ⇨, for example, *ISO reference model for open systems architecture*.

application program In any computer system, a program (or set of programs) which performs a specific task for end users, but not a program concerned with controlling the *computer* or parts of its *operating system*. Similarly, in *data communications* an *application program* is one which works upon *text* transmitted by the system, and has no responsibility for system functions such as *error checking*, *dialogue management*, *addressing*, *routing*, and *link control*.

architecture The architecture of a system relates to its design, and the way in which the component parts interrelate. The term can be applied to the design of any machine, such as a *minicomputer*, or to an overall communications network. The architecture refers to the logical structure of a system, rather than the specification details of the individual components used to construct the system. ⇨ *open systems architecture*.

ARQ ⇨ *automatic repeat request*.

articulation test A test made for the quality of line transmission in a *telephone* network. These tests are usually conducted to show the variation of sound articulation against *attenuation*, comparing an object system with a standard reference system.

artificial ear A device used to permit the calibration of earphones in telephone transmission systems, and having an acoustic impedance the same as the average human ear.

ASCII A version of the *International Alphabet No. 5* code, which has minor variations to cater for requirements in the USA. The term is pronounced 'askey' and is a contraction of *USA Standard Code for Information Interchange*. It is a 7-*bit* code giving 128 unique combinations or *characters*, of which 32 are reserved for special control functions. ⇨ *data communication code*.

aspect ratio The physical proportions of a viewing screen in a *television receiver* or other *visual display* device, expressed as the ratio of width to height. The usual standard is 4:3.

ASR Abbreviation of *automatic send and receive*.

assembler A *hardware* or *software* device which translates statements from one form to another; e.g. operates upon statements written in a programming language to produce a machine language program which can run on a particular processor.

associated channel signalling ⇨ *channel associated signalling*.

asymmetrical duplex transmission A system in which transmission takes place simultaneously in both directions over a *circuit* but not at the same rate in each direction.

asynchronous operation A method of *data transmission* in which each *character* is framed by *bits* or pulses (e.g. *start and stop signals*). The *start bit* triggers a timing mechanism in the *receiving terminal* which counts off succeeding bits of the character as a series of fixed time intervals. The *stop bit* resets the receiver ready for the next character. This technique is usually associated with slow speed devices like *teleprinters*. Contrast with *synchronous operation*.

asynchronous system A communications system in which units of *data* (e.g. *characters*, *words* or *blocks*) are preceded and followed by *start/stop signals* which provide timing at the *receiving terminal*.

asynchronous terminal A transmitting/receiving device which operates using *asynchronous signals*; i.e. in which units of *data* are bounded by *start or stop signals*.

attenuation An undesirable condition in which the *message signals* transmitted along a *channel* are corrupted due to the waveform of the *signal* being chopped off. Attenuation is related to the frequency of

the message signal and the *transmission medium* used in the link.

The attenuation attributed to any part of a system is usually measured in *decibels* (*dB*). For example, in a pair of wires attenuation is measured in dB per unit length, and is related to the resistance (R), the inductance (L), and the capacitance (C) between the pair per unit length.

To overcome the effects of attenuation, it is normal practice to use *repeaters*, which also deal with other signal impairments, and regenerate the signal to maintain a high *signal-to-noise ratio*.

In planning communication networks, it is necessary to consider the *nominal total attenuation* resulting from different *routing* situations.

audio amplifier A device used to increase the magnitude of a *signal* which is in the range of frequencies audible to the human ear.

audio circuit A *circuit* designed to carry a *sound programme* suitable for a radio broadcast or the sound component of a television broadcast. The *bandwidth* of the circuit would therefore allow a range of frequencies suitable for the representation of speech and music (i.e. in the range from 20 Hz to 20 kHz). A circuit designed for telephone *traffic* would be referred to as a *speech circuit* or *voice circuit*.

audio frequency ⇨ *audio frequency band*.

audio frequency band The spectrum of frequencies associated with the range of *sound waves* which humans can perceive; i.e. within the band from 20 Hz to 20 kHz. In communication systems it is necessary to distinguish between *transmission paths* able to cover the full range of such sounds, and more limited media used to carry only speech, as in *telephony*.

Speech circuits in telephony only require a *bandwidth* from 300 to 3,300 Hz, whereas *audio circuits* able to carry music and speech to the level of quality expected in radio and television broadcasts may cover the full audio frequency band.

audio frequency waveband The band of electromagnetic frequencies which equate to *sound* as perceived by humans. Same as *audio frequency band*.

audio signal An *electrical signal* corresponding to *sound waves*. The electrical signal represents the sound pressure fluctuations, and a direct means of conversion from sound to audio can be achieved by a *microphone*, and conversely from audio to sound by a *loudspeaker*. The audio signal is an electrical *analog* of sound.

automatic alternative routing A facility which enables a *call* which is blocked by busy *lines* on a *primary route* to be automatically diverted to an *alternative route*.

automatic announcements A facility provided in modern telephone *networks* in which digitally recorded segments of speech are delivered to *users* in specific situations; e.g. early morning calls, speaking clocks, and advice to callers on avoidance of incorrect actions.

automatic answering A facility in which a *terminal* automatically responds to a *calling signal*, and the *call* may be established even though the *called terminal* is unattended.

automatic calling A facility which enables a device to automatically *call* another *terminal* over a *public data network* by generating an *address* in the *calling terminal*.

automatic dialler A device which, in response to a given input *signal*, will automatically call a distant *terminal* over a *network*.

automatic equalizer ⇨ *equalization*.

automatic gain control A unit designed to compensate for variations in the strength of a *signal* received by radio propagation, or along a *transmission path*. It includes features to generate a control voltage which will increase the *gain* of an *amplifier* to maintain a more or less constant output, despite fluctuations in the power of the received signal.

automatic recall A facility whereby a *terminal* can automatically attempt to callback a terminal which is busy so that the desired *call* is established when the *called terminal* is free.

automatic repeat attempt In a *telephone* or *data network*, a facility provided to automatically repeat an attempt to establish a *call* when difficulty has been experienced in establishing the call. The repeat attempt may use the same *circuit* or select another circuit.

automatic repeat request (ARQ) A method of *error correction* in *data transmission*, in which any *blocks* of *data* found to contain errors at the *receiving terminal* are automatically requested for repeat transmission from the *transmitting terminal*. ⇨ *half duplex error protocol* and *full duplex error protocol*.

automatic send and receive (ASR) Pertaining to a *terminal* which has *transmitting* and *receiving stations* and logical facilities which enable incoming or outgoing *messages* to be stored (e.g. on *paper tape*, *magnetic tape*, or in *computer* memory).

automatic sequential connection A facility provided in *private* and *public data networks*, in which a particular *terminal* can call other terminals at specified *addresses* in a predetermined sequence. For example, a *computer* dialling *remote batch terminals* to collect batches of input data from distant locations.

automatic service Any service such as the public *telephone system*, in which a caller can make a necessary connection without the intervention of an operator at an *exchange*.

17

automatic switching equipment That part of an *exchange* in a *switched system* which *routes* calls automatically to make the connections requested by *calling parties*.

automatic transfer A special procedure used to transfer an incoming *call* from one *exchange* to another, or from one subscriber *terminal* to another.

availability of service The design of any communication system includes an attempt to minimize the use of resources (i.e. reduce system costs), while maintaining a high level of *user* satisfaction. One criterion by which the quality of service is measured is the availability of the system; e.g. whether a user *terminal* can gain access to the network without unreasonable delay. The performance of the system in this respect is expressed as a target probability of a *call* being blocked. For example, the target *probability for call blocking* on a *local exchange* might be 0·02. ⇨ *quality of service* and *reliability of service*.

average call duration A statistic used in *traffic theory*, and obtained by dividing the total number of minutes of conversation by all *calls*, by the number of *effective calls* in a given period.

average operating time A statistic used in expressing the quality of service of a *telephone system*, in which the total operating time in minutes for the period under review is divided by the number of *effective calls* recorded in the period.

average traffic The mean volume of *traffic* (E) handled in a given period of time by a particular part of a *telecommunications* system, and expressed as:

$$E = \frac{nh}{T}$$

where n = number of *calls*
   h = the mean holding time per call
and   T is the duration of the period.
⇨ *traffic theory* and *mean holding time*.

awaiting digits state A state which exists when a *line* is activated but awaiting *signals* from a *terminal* about to make a *call* over a telephone *network*.

B

backbone routing A *routing* which uses only the *main trunk* routes in an *exchange hierarchy* is known as backbone routing.

backing storage ⇨ *backing store*.

backing store That part of a *computer* store which contains programs and *data* not immediately needed for processing. Usually, backing store is on magnetic disc or *magnetic tape* and access can be obtained to segments of *data* which are available in amounts of several thousand *bytes*. Access time may take several milliseconds whereas, once in *main memory*, data can be accessed as discrete operands for processing in access time measured in nanoseconds.

back-off A procedure whereby the load is systematically reduced from an overloaded *channel* to prevent *collision*, or to *restart* in a *peak load* situation. This technique is used in *local area networks*.

back space (BS) A function represented as a special *format effector* in a *data communications code* and serving as an instruction to move a *print mechanism* or a *cursor* of a *visual display unit* backwards for one position.

backward channel A *channel* in which transmission takes place in the reverse direction to the intended *data flow*; the backward channel is normally used for supervisory or *error control* signals concerned with the main data flow over the *forward channel*. Data communication systems can transfer information in both directions and it is, therefore, necessary to relate this term to the *data source* at any instant.

backward direction Relating to the transmission of supervisory and *error control* signals in a *data network*, which take place on a *backward channel*; i.e. in the reverse direction to that in which information is being transferred.

badge reader A *data terminal* used for *data collection* in which the *user* inserts a badge (or plastic card) to initiate a transaction. The badge contains information to identify the user and may have authorization codes which delimit the nature of the transaction that can be performed.

balanced double-current interchange circuit An *interface* circuit between a *data terminal* (*DTE*) and a *modem*, having certain electrical characteristics as defined in the *CCITT* recommendation V11 concerned with *data transmission* over telephone *circuits*.

band pass filter A *filter* which allows a band of frequencies to pass along a *circuit*, whilst blocking all frequencies above and below the band. Contrast with *high pass filter* and *low pass filter*.

bandwidth The bandwidth of a *communication channel* defines the range of frequencies which can safely be conveyed in the channel. If an attempt is made to transmit signals outside the nominated bandwidth the signals may be distorted, such that information carried is lost or corrupted. For example, in a simple *intercom* system the objective is to communicate *voice signals* between two *users*, and for this purpose a bandwidth of 20 Hz to 3 kHz is sufficient. This will cover the *dynamic range* of sounds made in normal conversation and enable recognition of familiar voices. It will not provide a high fidelity reproduction of sounds occurring at each *station*.

In a *telephone system*, the restriction of

the allowed bandwidth can reduce the effect of *noise* at frequencies where there is little information transmitted, and a planned limitation of the bandwidth reduces the cost necessary for *amplifiers* and transmission links in the system, whilst retaining a *signal* acceptable to the users.

For public telephone systems, the *CCITT* recommendation for the *voice band* is 0·3 to 3·4 kHz. This can be compared with the bandwidth of 15 Hz to 20 kHz which is necessary for high quality music transmission systems.

The *video signal* of a television broadcast depends upon the number of *lines* used in the service, but, for example, the UK 625-line system requires a bandwidth of 5·5 MHz. In practice, broadcast signals are conveyed in channels of 8 MHz which include a separate band for the *sound signal* and additional bandwidth necessary for the *vestigial sidebands* which arise from the necessity to modulate the video signal on to a *radio frequency carrier* wave. ⇨ *modulation.*

A circuit which is capable of carrying only *low frequency* signals, e.g. *audio frequency* signals, is said to be of *narrow bandwidth*. A circuit which is capable of carrying *high frequency* signals, e.g. a *radio frequency signal*, is said to be of *broad bandwidth*. Very often a *broadband channel* is used to carry several narrow band channels at the same time by modulating the narrow band channels on to a high frequency carrier.

The electrical properties of the *transmission medium* have a great effect on the bandwidth of a particular *transmission channel*. Serious *attenuation* or *crosstalk* affecting the *message signal* will arise if the frequency characteristics of the signal are not matched to the properties of the medium. These problems can be overcome with equipment known as *repeaters* spaced at intervals along the transmission channel. However, this is expensive, and research to find improvements in transmission medium is important for the growth of the industry and the economics of communications.

The table shown below gives an indication of the bandwidth potential of different transmission media. It assumes that appropriate repeaters are installed.

Medium	Practical Frequency Band
twisted wire pairs	up to 500 kHz
co-axial cable	60 kHz to 60 MHz
waveguides	2 GHz to 11 GHz

Digital signals can be transmitted over a variety of transmission channels of different bandwidths. The digital signals exist usually at two discrete levels (e.g. corresponding to the *binary notation* 0 and 1). In the example shown in Appendix 6, two *bits* of information are represented by a signal inversion corresponding to a single cycle of an AC signal. The higher the channel bandwidth, therefore, the greater possible inversion rate and therefore the greater possible speed of information transfer.

The terminal equipment in a digital system is built to send and receive message signals consisting of pulses of a defined duration. The duration of the pulses determines the amount of information which can be received by the terminal in a second, and devices are rated in terms of pulses or bits per second, known as the *baud rate*. Some recommended rates of data transmission by the CCITT are 200, 600 and 1200 bauds.

Computer-to-computer links often require several thousand bits (kilobits) to be transferred per second and, in some cases, millions of bits (megabits) per second. For this purpose a broadband channel allowing high frequency transmission is needed. It is common in data transmission to use the term *bit rate* as being synonymous with channel bandwidth.

baseband co-axial system A communications system, suitable for a *local area network*, in which information is directly encoded on to a *transmission medium* consisting of a *co-axial cable*.

baseband modem A modem which has a more limited technical specification than is required for connecting a *data terminal (DTE)* to a *public switched telephone network*. A baseband modem can be used on *private lines* and for limited distance communication. Also known as a *data service unit*.

BASIC *A programming language* classed as a *high level language* and used for developing programs to run on a particular computer *hardware*. The name is derived from: Beginners' All-purpose Symbolic Instruction Code.

basic routing The *routing* involving the least number of *trunk links* is known as the basic routing.

batch processing A form of *data processing* in which transactions are collected in batches to be processed at a convenient time. A batch processing system may provide a useful information system (e.g. for accountancy purposes). It would not, however, be used to monitor and control fast-moving events in a real world, because the delays in collecting, preparing and processing batches may result in *files* being hours, or days, behind in recording the status of real events.
Compare with *real-time system*.

batch systems Any *data processing system* which uses *batch processing* techniques.

baud A term used to express the rating of equipment or a *transmission channel* in a communications system. The number of pulses which can be transmitted in a second is the *baud rate*. Baud translates as pulses per second, just as *hertz* equals cycles per second. The definition is named after Emile Baudot, an early contributor in the development of *telecommunications*. ⇨

modulation rate, and compare with *data signalling rate*.

baud rate The speed of operation of a device or *channel* in pulses per second. Also known as *modulation rate*, and should be compared with *data signalling rate*.

Baudot ⇨ *Emile Baudot*.

Bel A unit of measure for the overall *loss* or *gain* in power attributable to a *circuit* or device. The loss or gain is given by the formula:

$$N = \log_{10} \frac{P_2}{P_1} \text{ Bels}$$

where N = the number of Bels
P_1 = power sent
P_2 = power received

In practice, it is more convenient to use the measure *decibel* which is one tenth of a Bel.

BEL A *signal* forming part of a *data communication code* and used to activate an audible alarm at a *receiving terminal*. An abbreviation of bell and used to attract an operator's attention.

bi-directional communication A form of communication in which two *users* are able to send and receive information over a *network*. This can be achieved by a *full duplex* operation in which both parties can send and receive simultaneously, or by a rapid exchange of information, in which the parties alternately send and receive information.

bid In any *switched system* an attempt, successful or otherwise, to secure a *circuit* in a *circuit group*.

bids per circuit per hour (BCH) A ratio intended to indicate to operators of a switched network the *traffic pressure* for a group of *circuits*. The ratio is given as the number of *bids* per hour, divided by the number of working circuits.

bilateral control Relating to the synchronization of *nodes* at different locations, in which the *clock* at each node can control the clock at the other; i.e. one node is not a slave to the other.

Bildschirmtext The name given by the Deutsche Bundespost to its *public videotex* service, introduced into West Germany as a trial starting in 1980. ⇨ *videotex system.*

billing The process concerned with preparing invoices which are intended to charge *users* for their utilization of a communications system. This includes the procedures concerned with measuring their use of the *network.*

billing information Information collected in order to produce charges to subscribers for use of a communications service.

binary coded information Information which has been represented as *binary numbers*, or patterns of *binary digits*, in order that the information may be transmitted or processed.

binary digit A *digit* in *binary notation*, i.e. 0 or 1. A group of binary digits may represent a number (e.g. 011011 in binary represents 27 in decimal form), or they may be used in groups to form binary codes which represent alphabetic *characters*, numerals or special symbols. As an example, it is common in *computers* and other electronic systems to use 6-bit groups to form up to 64 unique patterns which represent 64 separate characters. An example of such a *code* used commonly in *data transmission* is shown in Appendix 1.

binary digital signal A *signal* in which information is represented as *binary numbers*; e.g. is represented by two voltage levels in a *circuit.*

binary notation A system for representing numbers in which each digit position may contain either an integer 1 or 0 (i.e. in

a radix of two). In the addition of integers, the following rules apply:

$$
\begin{array}{ccc}
0 & 0 & 1 \\
+0 & +1 & +1 \\
\hline
0 & 1 & 10
\end{array}
$$

Numbers in binary notation thus appear as rows of 1's or 0's, and the range of values which can be represented increases with the number of positions. For example, four digit positions give 16 unique patterns of *bits* including:

$$0000 = 0 \text{ in decimal notation}$$
$$0001 = 1 \text{ in decimal notation}$$
$$0010 = 2 \text{ in decimal notation}$$
$$1110 = 14 \text{ in decimal notation}$$
$$1111 = 15 \text{ in decimal notation}$$

Inside a *digital computer*, or any other digital machine, binary information is easily handled since each integer can be represented by a high or low voltage.

binary number A number in which the radix for each *digit* position is two, and the numbers are represented by the digits 0 and 1.

binary string A succession of contiguous electrical pulses representing *binary coded information* and treated as a unit of *data* for processing or transmission purposes, but not necessarily a complete *data element* or set of data elements intended for an end *user.*

binary tariff system A system of charge made up of two elements. For example, in the operation of a *telegraph network* a fixed charge is made for the acceptance and delivery of a telegram, and a further variable charge is made, depending upon the length of the telegram in words.

bipolar transmission A technique used in the transmission of *digital signals* to reduce the apparent effect of a direct *current* (*DC*) component in the *binary signal*. This effect arises because the number of 0's and 1's is not equal and, on average, this creates the effect of a DC element in the signal. In bipolar transmission, a 1 is represented by either a positive or negative voltage, and a zero by 0 volts. The positive and negative values are used alternately, and therefore the average value of the signal is always zero.

bit Abbreviation of *binary digit*.

bit-clocking The process of maintaining *synchronization*, say between a *terminal* and a *network*, by the transmission of *bits* (binary digits) which provide co-ordinating timing pulses for various *control* and transmission activities.

bit error In digital systems, *data* is transmitted as a series of discrete electric pulses at levels representing the 1 or 0 *digits* of a *binary number*. It is possible for intermittent errors to arise which cause *bits* (binary digits) to be received incorrectly. These are known as bit errors. Various *error checking* procedures are used to detect and correct bit errors, including the automatic retransmission of *frames* known to contain errors. ⇨ *error checking* and *error rate*.

bit error rate A measure of the quality of a *circuit* used for *data transmission*, expressed as a ratio of the number of *bits* incorrectly received to the total number of bits transmitted (e.g. 1 in 10^7 bits).

bit-order of transmission Describes the arrangement for the transmission of any unit of *data* in a system, where *serial transmission* is used. For example, the most significant *digit* of a number or *field* may be sent first, or the least significant digit first.

bit parallel transmission ⇨ *parallel transmission*.

bit pattern A specific pattern of *binary digits* used in a system to represent a particular instruction or meaning.

bit rates Pertaining to the speed of operation of digital equipment. For example, a 9·6 *kilobit* channel can handle *digital signals* at speeds of 9,600 *binary digits* per second. ⇨ *modulation rate* and *data signalling rate*.

bit serial transmission ⇨ *serial transmission*.

bit-stream *Bit* is an abbreviation of *binary digit* and is used to describe a single element used to represent 0 or 1 in a *binary number*. In modern *telecommunications*, many of the techniques associated with *digital computers* have been applied; thus, both *text* (i.e. user information) and *control information* are transmitted as streams of binary digits or bit-streams.

bit stuffing In *time division multiplexed systems* it is necessary to synchronize *time slots* in *switching* operations. To ensure that one *channel* coincides with the *clock* of another, the technique of bit stuffing is sometimes used. This entails the insertion of redundant *bits* into an incoming *bit-stream* to increase the rate; the presence of the stuffed bits is signalled to the receiver so that they can be removed to restore the original *data*. ⇨ *time shifting*.

bit synchronization Whenever digital devices communicate over a *network*, the individual pulses which represent *bits* (*binary digits*) have to be synchronized by an electronic *clock* which maintains the timing sequence required between two communicating devices to ensure that bits are not lost. Such clocks in *synchronous digital systems* are in the *data terminal* or the *modem* of the *transmitting station*. The signals sent create transitions on the *line*, and the clock at the *receiving station* adjusts its clock rate to these transitions. Where long high-speed *digital transmissions* take place, special synchronizing characters (*SYN*) are inserted into the data at the beginning of transmission, to establish *synchronization*. ⇨ *character synchronization* and *message synchronization*.

23

blackbox An expression to refer to any item of equipment which carries out a specific set of functions within a system. Usually used to simplify a discussion of an overall system; for example, between technician and layman, in which the significant objective is to convey the idea of the functions, and inputs and outputs required by the blackbox.

block A group of *bits*, *characters* or *words*, transmitted as a unit of *data* in a system and over which any *data link control* procedures may be applied to effect *error detection* and/or *error correction*.

block-acknowledged counter A facility provided in a *terminal* which is transmitting *blocks* of *data*, to record the number of blocks which have been acknowledged as received by a distant *receiving terminal*. Compare with *block-completed counter*.

block check A check performed in a *data network* to facilitate *error control* and usually concerned with predetermined rules for the formation of blocks.

block-completed counter A facility provided in a *terminal* which is transmitting *blocks* of *data*, to record the number of blocks which have been transmitted. Compare with *block-acknowledged counter*.

block error rate In a *data network*, a measure of the quality of a *circuit* used for transmission, expressed as a ratio of the number of *blocks* incorrectly received to the number of blocks sent.

block separator A *character* which defines the format of information structured in *blocks*, and used to indicate that the next character in a sequence belongs to a new block of information.

blocked call A *call* is said to be blocked when *access paths* from the *called terminal* to the *calling terminal* are fully engaged with existing *traffic*.

blocking signal

(1) A *signal* sent to an *exchange* (or *node*) to indicate that the exchange should not use a particular *circuit* for any *outgoing call*. The exchange is, however, able to receive an *incoming call* on that circuit. This signal is used in maintenance procedures.

(2) Any signal sent on an idle circuit to prevent the circuit being seized by another exchange or user *terminal*.

blocking signal acknowledgement A *signal* sent by an *exchange* to acknowledge that a particular *circuit* has been blocked in response to a *blocking signal*.

bounded medium ⇨ *bounded transmission medium*.

bounded transmission medium A physical set of materials used to carry *signals* is known as a *transmission medium*. A *coaxial cable* is an example of a bounded transmission medium; i.e. the signal is propagated along a physical path determined by the properties of the medium. By contrast, the atmosphere through which *radio signals* are propagated is known as an *unbounded medium*.

bps Abbreviation of *bits* per second; a measure of the rate of operation of a *circuit* or device.

British Telecom A public corporation responsible for the operation of public *telecommunication* services throughout the United Kingdom of Great Britain and Northern Ireland; including *telephone*, *telex*, *videotex*, and *public data networks*. The corporation was given a separate identity during 1981, prior to which it had operated as the Telecommunications Division of the British Post Office.

broad bandwidth ⇨ *bandwidth*.

broadband channel A *transmission path* having a wide *bandwidth*. For example, a *television channel* requires a bandwidth of approximately 5·5. MHz as compared with a telephone *speech channel* which affords only 3 kHz. A broadband channel would be constructed of materials which do not offer significant resistance, *attenuation*, or *distortion* of high frequency *electromagnetic waves.* ⇨ *bandwidth.*

broadband multiplexing channels A *channel* which has a sufficiently high *bandwidth* to enable a number of *message signals* to be combined and transmitted over the same channel, using *multiplexing* techniques. For example, a 2000 MHz channel can accommodate 500,000 *telephone channels* where each telephone channel is allowed 4 kHz. ⇨ *multiplexing.*

broadband transmission channel ⇨ *broadband channel.*

broadcast Any form of transmission in which all *subscribers* connected to a particular service are addressed at the same instant with the same *message signal.* The *channel* for communication could be a *bounded medium* such as a *cable network*, or an *unbounded medium* such as a television signal broadcast as *radio waves* in the atmosphere.

broadcasting organization Any organization which is concerned with *sound* and/or vision broadcasting; e.g. providing radio, television or *teletext* facilities.

BS ⇨ *backspace.*

buffer A storage unit used to retain a unit of *data* until it can be processed by another device; e.g. to balance the capacity of a high-speed *digital computer* with the relatively slower speed of a *transmission channel.*

buffered terminal Any *terminal* which has a magnetic memory device which can be used to store incoming or outgoing *messages.* For example, any terminal, which is activated remotely by the *polling* action of a *computer*, must have the ability to

store information until its turn in the polling sequence arises.

buffering A technique used in *data transmission systems* to balance the *traffic* to the capacity of some part of the system. For example, incoming *messages* are held in a *buffer* until they can be serviced, and outgoing messages until they can be transferred along the *transmission line.*

burst errors Errors occurring in a *circuit* used for *data transmission*, in which the frequency of the errors is such that less than a specified number of *correct bits* occurs between *error bits.* ⇨ *transmission errors.*

burst isochronous transmission The transmission of *synchronous data* in bursts over a *public data network* to produce a mean *data signalling rate* on an *information bearer channel*, compatible with the *input data signalling rate* of a receiving device.

bursty traffic Refers to *traffic* which arises in bursts and, in particular, to *data networks* in which the traffic level may change in a short time-period from very low to very high, approaching the *peak volume* planned for the *network.*

business terminal A *data terminal* used in a business environment, or a telephone *terminal* used in a business rather than a residential environment.

busy hour A period of uninterrupted time (nominally 1 hour), for which the *traffic* in a *telecommunications* system is at its maximum level.

busy hour average traffic The mean *traffic volume* arising in the period of peak utilization of a *telecommunications* system.

busy hour traffic In planning the capacity required in a communications system, it is necessary to assess the *traffic volume* expected during the busiest period. For example, in a *public switched telephone net-*

work, the traffic tends to reach a peak during the morning of a normal working week-day. An hour is the period chosen because a shorter period would make demarcation of the period difficult, whereas a longer duration would tend to average out peaks to be considered.

busy tone The *signal* heard by the *user* of a *telephone* when the *subscriber* being called has already lifted the *handset* to speak to another party. Also known as *engaged tone*.

byte A unit of *data* used in many modern *computer* systems and also used as a unit

of *data transfer*. A byte usually consists of two *characters* or 8 *information bits*.

byte-order of transmission Pertaining to the sequence in which the successive *bytes* which make up a unit of *data* are transmitted. For example, the most significant byte may be transmitted first, or the least significant byte first.

byte-serial transmission A method of transmission in which successive bytes are transferred serially, in the appropriate sequence of the *data*. It should be noted that the individual *bits* (*binary digits*) that make up each *byte* may not necessarily be transmitted serially.

C

cable loop A cable used for transmission in a *local network*, arranged in a loop to which all devices are connected for communication. Thus, all devices share the same *channel* and *contention* may arise.

cable network A *network* for *data communication* or for the transmission of television or radio broadcast signals. Usually refers to a network of *co-axial tubes* and, in the case of broadcast transmission systems, to provide a closed-circuit one-way path for the delivery of *video* and *sound* signals.

cable pressurization A technique used to protect cables in underground cable ducts, by pumping dry compressed air into the cable itself. This tends to reduce damage caused by the ingress of water and thus reduces faults on cables; it serves also to identify cable damage by a loss of pressure.

call A term used to describe the process of communication between *users*, and more particularly to refer to the process which takes place in communication over the *public switched telephone network* or *public data networks*. This process can be considered in three phases, *call establishment* (making the initial connection between *terminals*); *conversation* or *data transfer* (transferring *messages* between user terminals); and *call clearing* (the orderly disengagement of terminals at the completion of a *call*).

call accepted packet A response given to the *network* by a *terminal* in a *packet switching system*, when it is ready to accept a *call request* from another terminal. This action allows the network to establish a *virtual circuit* between the *calling terminal* and *called terminal* so that *data transfer* can take place.

call accepted signal A control *signal* sent by a *terminal* to indicate that it has accepted an *incoming call* on a *data network*.

call accounting system A subsystem in an *exchange* which collects and processes charging information in respect of *calls* made by *subscribers*.

call answered signal A *signal* which arises when a *called terminal* responds to a *call*; e.g. when a *called subscriber* lifts the *handset* of a *telephone* to respond to a call. This signal establishes a *speech path* between the two terminals and usually starts a *billing procedure*.

call-back when busy terminal becomes free This facility can be invoked by a *calling terminal* when it has received a busy *signal* from a terminal to which a *call* has been requested. The facility will act as an instruction to the *control* system to establish the desired call when the busy terminal becomes free.

call blocking If all the *channels* in a particular *trunk route* are in use, then any new *call* arising will be *blocked*; i.e. the requested connection will not be accepted and an engaged tone may be sent.

Telecommunication systems are usually planned to present a low probability of call blocking. The probability is expressed as:

$$B = \frac{C_B}{C_O}$$

where C_O = calls offered
C_B = calls blocked

In practice, the probability of blocking in the *busy hour* is targeted to be around 0·01. It should be seen that the *average traffic* must be less than the number of channels available to provide such a performance.

The probability of call blocking is an

important measure of the performance of a *circuit switched* system. In *message switched* systems, the equivalent performance of the system is measured in terms of the *messages* taking more than a specified time to reach their destination. ⇨ *traffic theory*.

call clear-down time The time taken to clear a *call*, from the moment the action is initiated by a *terminal*, to the moment a free condition is signalled on the *data terminal* originating the call. Same as *call release time*.

call clearing The activity associated with making a *call* on a *switched circuit* is usually considered in three phases. These are: *call establishment*, *message transfer* and *call clearing*.

Call clearing covers the procedures and the activity associated with the correct and orderly disengagement of the two *terminals* at the completion of the message transfer.

The concept applies to both switched circuits for *digital transmission* and *voice telephony*, but the detailed procedures are different.

call connected packet A control *packet* transmitted in a *packet switching* network to signify the establishment of a *virtual circuit*. ⇨ *packet switching*.

call connected signal A control *signal* in a *circuit switching* system which signifies to a *calling terminal* that a connection has been completed in response to its *call request*.

call control procedure The set of actions and *signals* which are required to establish, maintain and release a *call* in a *network*.

call control signals The complete set of *signals* which are required to establish, maintain and release a *call* in a *network*.

call disestablishment The procedures concerned with the orderly termination of a *call* at the completion of the phase in which information has been transferred, and concerned with the release of *circuits* and resources which have been allocated to the call.

call duration For a given *call*, the period of time that elapses from the moment when the *call accepted signal* is received by the control system for the *network*, until the *clear forward signal* (or *clear backward signal*) is received as a result of a *terminal* clearing the connection.

call, duration, average ⇨ *average call duration*.

call establishment The activity associated with making a *call* on a *switched circuit* is usually considered in three phases. These are: call establishment, *message transfer* and *call clearing*. Call establishment covers the procedures and the activity concerned with the *calling terminal* making a connection with the *called terminal*.

The concept applies to both switched circuits for *digital transmission* and *voice telephony*, but the detailed procedures are different.

call failure signal A *signal* transmitted on a *backward channel* to notify a *calling terminal* that its *call request* cannot be completed; e.g. because an event in the *call control procedure* has not been completed in the specified time.

call not accepted signal A *signal* sent by a *data terminal* to indicate that it will not accept an *incoming call*.

call processing system A subsystem in a *digital exchange* which controls the progress of each *call* on the basis of instructions received from *users*.

call progress signal A *signal* between *data circuit terminating equipment* (*DCE*) and a *data terminal* (*DTE*) to notify the DTE of the progress in making a *call* which has been requested. The call progress signal may indicate positive or negative progress.

call release time ⇨ *call clear-down time.*

call request ⇨ *call request signal.*

call request packet In a *packet switching system*, the *calling terminal* delivers a *call request packet* to the *network* when it requires to establish a *virtual circuit* to another terminal.

If a *circuit* is available, and if the called terminal is able to accept the *call*, a *call accepted packet* is delivered by the network to the calling terminal and the *data transfer phase* is started. When this phase is completed the calling terminal instructs the network to deactivate the link by a *clear request*.

call request signal A control *signal* requesting service from a *network* sent by a *calling terminal* to the network, in which the *address* of the terminal to be called is indicated. In the case of a *telephone*, the *call request* is indicated by lifting the *handset* and dialling the required address.

In *data communication systems*, the call request may include other information sent as a pattern of *bits* to the *line*.

call set-up The process of establishing a link between two *terminals*, including the identification of the *address* required by the *calling terminal*, the selection of a path through the *network*, and the acceptance of the connection by the *called terminal*.

call set-up time The overall length of time required to establish a *call* between two *terminals*, starting from the time for the initiation of the *calling signals*, until a *call connected signal* is delivered to the terminal originating the call.

called line identification A system in which the *network* confirms to the *calling terminal* the identity of the terminal with whom a connection is about to be made, thus allowing the connection to be cancelled. ⇨ *calling line identification.*

called location The *terminal* addressed as the destination for a particular *call* or *message signal*.

called party The person who is to receive a particular *call* or *message signal* at a *called location*.

called subscriber The *user* to whom a *message* is sent, or the user who is addressed by a *call request* originated by a *calling subscriber*. Also known as *called party*.

called terminal A *terminal* which has been identified by a *call request* issued by another terminal wishing to transmit a *message* or conduct a *conversation*. Contrast with *calling terminal*.

called terminal alerted state A *state* that exists when a *subscriber* in a *telephone system* has completed dialling another *terminal* which is available to receive a *call* and the call bell is being rung.

called terminal answered signal A *signal* sent in the *backward direction*, indicating that a *called terminal* has answered and *call establishment* is in progress.

called terminal engaged signal A *signal* sent back to a *calling location* to indicate that the *called terminal* is engaged in another *call*.

called terminal free signal A *signal* (e.g. ringing signal) which indicates to the *calling terminal* that the *called terminal* is available but has not yet answered.

calling The action involved in making connections between *subscribers* to a switched network. ⇨ *abbreviated address calling, automatic calling, manual calling* and *multi-address calling*.

calling indicator signal On a *switched data circuit*, this is an *interchange signal* between a *modem* (or *DCE*), and a *data terminal* (*DTE*). The calling indicator signal is switched on when the DCE receives the ringing signal from a *calling terminal*. If the

DTE is switched on and in operable condition, it responds to the DCE with a *data terminal ready* signal which, in turn, causes the DCE to respond to the calling terminal with a *data set ready* signal.

calling line identification A system in which *called subscribers* are advised of the caller before *message transfer* can take place, thus allowing the *called terminal* to accept or reject the connection. Sometimes used in *data communication* to regulate priorities. ⇨ *called line identification*.

calling location The *terminal* which originates a *call* or *message signal*.

calling party ⇨ *calling subscriber*.

calling party clear A method of *call clearing* in which the *call* is not cleared until the *calling party* puts down his *handset*. If the *called party* alone puts down the handset, it can be lifted to continue the call. Contrast with *first-party clearing*.

calling party's category indicator A special item of information attached to a *message* to notify the *called party* about the nature of the *call*. For example, indicating priority.

calling rate The measure of the use made by *users* of a specific *terminal* connected to a *telecommunications* system; i.e. a calling rate of 3 *calls* an hour of average duration of 4 minutes per call gives 12 minutes' utilization of the system per hour. Thus, the terminal originates 0·2 *erlangs* of *traffic* in one hour.

calling signals Coded information which instructs a *network* to establish a *route* to a particular subscriber *terminal* in a *telephone, telex* or *data network*. Also known as *address digits* or *selection digits*.

calling subscriber In a *network* conversation, the *user* who initiates a *call* is known as the calling subscriber (or *calling party*). The user who receives the call is known as the *called subscriber* (or *called party*).

calling terminal A *terminal* which has issued a *call request* to a *network* to identify that it wishes to send a *message* or conduct a *conversation* with another terminal. Contrast with *called terminal*.

calls barred A facility which prevents a *terminal* from making *outgoing calls* or receiving *incoming calls*. For example, to reserve a *circuit* for a planned event.

CAN ⇨ *cancel*.

cancel (CAN) A special *code* appearing in a *message* recorded in *binary coded* form, and giving a specific instruction to the *receiving station* to disregard the preceding *data* in the *message* or *block*.

capacitance The property of *conductors* to store an electrical charge. The capacitance is defined as the ratio of the electrical charge between two conductors and their potential difference (Q/V). This property can create limitations in the transmission of *signals* over *circuits*.

Captains The name given to a *videotex* trial started in Japan in 1980 under the authority of the Ministry of Post and Telecommunication (MPT) and the Nippon Telegraph Telephone Public Corporation (NTTPC). The name is an acronym for 'character and pattern telephone access information network system'. The system is noted for the techniques used to transmit and display the Kanji characters used in the Japanese language. This technique is based upon an alphageometric pattern generator. Katakana and Hiragana, the other forms of Japanese syllabary, are also handled. ⇨ *videotex system*.

carriage return (CR) A function represented as a special *format effector* in a *data communications code* and serving as an instruction to return a *print mechanism* or a *cursor* of a *visual display* unit backwards to the beginning of the same *line*.

carried traffic The volume of *traffic* accepted by a system as distinct from the demand offered to the system by the *users*. Contrast with *offered traffic*.

carrier
1. Same as *carrier wave* or *carrier signal*.
2. Sometimes used to describe a communications authority providing *circuits* to carry the private *traffic* of individuals or corporations. ⇨ *common carrier*.

carrier detector signal An *interchange signal* between a *modem* and its associated *data terminal* (*DTE*) indicating that the modem has responded to a distant modem and is about to accept *data*. The DTE will not accept data unless the carrier detector is on, and this is designed to prevent the DTE responding to line noise which might be acted upon as data. When the carrier detector is switched off, the *call* is terminated. ⇨ *request-to-send* and *ready-to-send*.

carrier sense signal In some implementations of *local area networks*, a number of *stations* may be connected to the same physical *transmission channel*. Due to the high speed of the *channel*, it is possible for stations to interchange *frames* of information along the channel by simply sharing the channel availability on a random basis. A carrier sense signal is automatically applied to the channel when at least one station is attempting transmission. This normally acts to instruct other stations to defer to the transmitting channel and to be ready to examine the frames being transmitted.

carrier signal A *signal* generated for the purpose of carrying another *message signal* at a particular point in a *frequency spectrum* and used in the process of *modulation*. Same as *carrier wave*.

carrier wave A *signal* generated for the purpose of carrying another *message signal* at a particular point in a *frequency spectrum* and used in the process of *modulation*. Same as *carrier signal*.

cathode ray tube (CRT) A device used to display information, in which the information to be displayed is input in electrical form, and is converted to light from a luminescent screen. Used in *television receivers*, *data terminals* and *radar* equipment.

CCIR Abbreviation of *Comité Consultatif International de Radiocommunication*. A committee established to promote standards for the development of radio communication. This committee is set up under the *ITU – International Telecommunications Union*.

CCITT Abbreviation of *Comité Consultatif International de Téléphonie et de Télégraphie*. An international committee established to promote standards for the development of *telephone*, *telegraph* *systems* and *data networks*, and to create the environment for interworking between the *networks* of the different countries of the world. This committee is set up under the *ITU – International Telecommunications Union*.

cellular radio A branch of telecommunications technology concerned with the use of radio transmission to support mobile telephone and or data transmission. It is ideally suited for use with mobile objects such as cars or boats, because it offers a cheap and effective method to provide high volume coverage of a territory. The term is derived from the fact that a territory is divided into smaller areas or cells known as transmission zones.

The size of a cell may vary but would typically be between 2 to 10 miles wide. A separate transmitter/receiver is established within each cell and handles signals related to all users within the cell. A scanning system assesses the progress of users from one cell to another by monitoring the signal level. A central computer automatically switches user control to the appropriate cell. Cells may be clustered under the control of a central computer, which can make access available to other clusters or external networks such as the *switched public telephone network*, or *public data networks*.

The system is designed to allow a large number of users to operate the same radio frequencies without mutual interference. The cellular telephone terminals can provide facilities for direct dialled international calls, and can be used for data transmission when equipped with an appropriate *modem*.

This system has been introduced into many countries during the mid-1980s and is expected to develop as a mass communication system by the end of the decade.

centralized control signalling A system in which *call control signals* relating to a group of *data transmission* circuits are transmitted over a dedicated *circuit*. The same as *common channel signalling*. Contrast with *channel associated signalling*.

centre A term used to define a *switching* facility as an *international switching centre*.

chain A series of *circuits* connected together by devices for a particular purpose. For example, an international chain is made up of 4-wire *international circuits* connected to other national or international *4-wire circuits*.

changeback The process of transferring *traffic* back to a regular *circuit* after it has been temporarily transferred to a *reserve link* for maintenance operations.

changed-number signal A *signal* sent automatically in the *backward direction* to indicate to a *calling terminal* that the number of the *called party* has been changed recently.

changeover The process of transferring *traffic* to a new *circuit* because the existing circuit is faulty or needs to be used for another purpose.

channel A *channel* is a link between two *terminals* over which the *users* at each end can communicate with one another.

The simplest form of channel might consist of a pair of wires connecting two *telephones*, but a channel can consist of a complex set of physical resources linked to make a particular *call* feasible. In *data communication*, a channel may be a one-way communication path providing a *go* or *return path* for a *circuit*.

Some channels may be referred to as *broadband channels*; i.e. they allow hundreds or thousands of calls to be passed simultaneously along a physical path, using the technique known as *multiplexing*.

A *multiplexed channel* is a single channel which occupies a particular *frequency band* or *time slot* in a *multiplexing* system and is used for the duration of a particular call.

In some systems (e.g. a *local loop* network) all terminals may share the same channel without any attempt to separate different *messages* by frequency.

channel associated signalling A method of *signalling* in which the *signals* needed to control *traffic* on a particular *channel* are carried on the channel itself, or in a channel permanently associated with the traffic channel.

Contrast with *centralized control signalling* and *common channel signalling*.

channel identification

1. In *data networks*, information is often transmitted in *time slots* which are designated to carry particular *messages* in the form of *digital signals*. The concept arises of several *logical channels* occurring over a single *physical channel*. Each logical channel has to be identified and a channel identifier is used for each designated channel.

2. In modern *signalling* systems, a specific channel is designated to carry signals which control the progress of *calls* on a large number of *transmission paths*. Each signal has to include information to identify the channel to which the signal relates. This information is termed channel identification.

character

1. A letter, numeral or special symbol

(e.g. A, 1, or $), forming part of a *data communications code*.

2. A group of *bits* representing a letter, numeral or a special symbol in *binary coded* form.

character, check ⇨ *check character*.

character check A part of an *error checking* procedure designed to ensure that *character codes* conform to a valid *bit pattern* for the formation of a *character*.

character code A unique pattern of bits representing a numeric, alphabetic character, or a punctuation mark or special symbol. A method for representation of such *characters* in a *data communications code*.

character error rate In a *data network*, a measure of the quality of a *circuit*, expressed as a ratio of the number of *characters* incorrectly received to the total number of characters sent.

character framing Relating to a method of *data transmission* in which *synchronization* is observed between the *transmitting* and *receiving stations* for the duration of a single *character* only, and not between characters. Each character is 'framed' by a *start code* and a *stop code*. ⇨ *character synchronization*.

character generator A device which creates *characters* for display on a screen of a *visual display* terminal. For example, in a *videotex terminal*, the characters to be displayed are received as *codes* from the telephone *line* and held in a *page store*. The character generator processes information in the page store to create a character in the form of a dot matrix to be displayed in appropriate position on the screen.

character-mode terminal An *asynchronous terminal* operating in start-stop mode (e.g. to the *X-28* operating standards of the *CCITT*), and covering a range of slow speed *data terminals* operating in the range,

say, from 100 to 2,000 *bits* per second. The method of operation is characterized by a method of transmission in which each *character* transmitted is framed by a *start bit* and *stop bit*.

character-order of transmission Pertaining to the sequence in which the successive *characters* which make up a unit of *data* are transmitted. For example, the most significant character may be transmitted first, or the least significant first.

character-serial transmission A method of transmission in which successive *characters* of information are transferred serially. It should be noted that the individual *bits* (*binary digits*) that make up each character may not necessarily be transmitted serially, but perhaps as a *parallel transmission*.

character set A group of *characters* (i.e. letters, numbers, punctuation marks or special symbols) and the code formats by means of which they are represented in electronic systems. ⇨ example in Appendix 1.

character signal A set of *signal* elements which represent a character in the particular form required for transmission. For example, in *pulse code modulation* the quantized value of a sample.

character synchronization Character synchronization is relevant to both *asynchronous systems* (also known as start-stop systems) and *synchronous systems*.

With asynchronous systems, synchronization is maintained for the duration of a single *character* only; each character starts with a *start code* and the subsequent *bits* (say, 8 per character) are counted off and then terminated by a *stop code*. The stop condition is maintained on the *line* until another character is to be transmitted, whereupon a further start code followed by an 8-bit character sequence occurs, and so on.

33

With high-speed synchronous *data transmission*, the start-stop method is considered inefficient and, instead, bits are transmitted as a continuous stream with an electronic *clock* at the *transmitting station* serving to maintain the rate of inversion on the line. The *receiving terminal* adjusts to this timing and counts off characters as groups of bits. Special synchronization characters (*S Y N*) are transmitted at the beginning of a transmission to allow the *receiving station* to adjust to this timing. ⇨ *bit synchronization* and *message synchronization*.

charge, answer signal ⇨ *answer signal, charge*.

chargeable duration The time interval upon which the charge for a *call* in a *public network* is based. In *telephone systems* there is often a minimum chargeable duration (e.g. a 3-minute charge), but beyond this minimum duration the *calling subscriber* pays per minute. In *telex* systems or *public data networks* based upon *switched systems*, the chargeable duration begins at the moment the *call* is established and ends when either party terminates the call; the *tariff* may be based upon small fractions of a minute.

check bit A *bit* associated with a *character* or *block* of *data*, and used for checking the presence of an error in the character or block. ⇨ *parity bit*.

check character A *character* generated by an arithmetic process which is performed upon a unit of *data*; the character itself is added to the data to provide the basis for a *redundancy checking* operation.

check digits A pattern of *binary digits* derived from a unit of *data* (generated by an arithmetic process) and to be appended to the data for performing a *redundancy checking* operation.

check loop A device which is connected across the *go* and *return paths* of a *circuit*

to enable a *loop test* to be made upon the circuit.

chrominance components The elements of a colour television picture which carry information related to the colour of the picture. The chromaticity of a colour source is measured independently of the *luminance*.

circuit A set of physical transmission resources (e.g. *lines* and *exchanges*) which provide for two-way transfer of *message signals* from source to destination in a *telecommunications* system. The term 'circuit' usually implies that there are two *channels*, one for the *go path* and one for the *return path*. Where a *data terminal* (*DTE*) is transmitting to a *line* via a *modem* (*DCE*), channel 1 usually modulates (i.e. transmits), and channel 2 demodulates (i.e. receives). However, since *data* can be exchanged by two terminals, the modem has to be able to receive on channel 1 when required.

This concept occurs in various forms of communication, and it is common to describe a circuit as having a *forward channel* over which data is transferred, and a *backward channel* over which control *signals* and supervisory information are transferred. The forward and backward designations are relative to the direction of transmission at any instant in time.

circuit access points Points in a *circuit* which are accessible to engineers to enable transmission measurements to be made.

circuit group A group of *circuits* established for some particular purpose; e.g. to provide international communication between two communication authorities. ⇨ *grouping*.

circuit group congestion signal A signal sent in the *backward direction* to indicate that a particular *call* is unable to secure a connection due to congestion of a *circuit group* which must carry the call.

circuit switched connection A *circuit* which is established by a switching centre in an *exchange*, upon request from a *terminal* wishing to communicate with another.

The two terminals have exclusive use of the *transmission path* until the connection is released. Contrast with *message switched system*.

circuit-switched data network A *data network* in which *transmission paths* are established by a *switching* operation to make connections between *terminals* for the duration of a *call*. *Data transmission* can take place only when connections are made from end to end. Contrast with *message switched system*.

circuit switched exchange An exchange in which *transmission paths* are created by making connections between *incoming lines* and *outgoing lines*. Once a *call* is established between two *users*, the connection remains available and dedicated to the call until it is terminated. ⇨ *switched telecommunications system* and *switching equipment*.

circuit switched system ⇨ *switched telecommunication system*.

circuit switching ⇨ *switched telecommunication system*.

clear A process or a *signal* associated with a *calling party* or a *called party* taking action to terminate a *call*.

In a *telephone network*, if the called party replaces the *handset* before the calling party, then a *clear-back signal* (also known as a *hang-up signal*) is initiated.

In contrast, if the calling party clears first, a *clear-forward signal* is initiated.

These signals occur similarly in *public data networks* as a result of instructions originating in a *data terminal* (*DTE*) which pass to the associated *data circuit terminating equipment* (*DCE*). The DCE in turn instructs an *exchange* to terminate the call.

Although the clearing procedures vary for different types of network, the principal actions remain the same: the circuits utilized for the call are released, the measurement of *call duration* is stopped, and charging is stopped.

clear back signal A *signal* sent in the *backward direction* to signify that the *called party* has terminated a *call*. Same as *clear-backward signal*.

clear-backward signal A control *signal* which terminates a *call* upon the instruction of the *called party*. Same as *clear back signal*.

clear confirmation A *call control signal* between a *terminal* and a *modem*, acknowledging a request to clear a *call*.

clear-forward signal A *signal* sent when a *calling party* has decided to terminate an established *call* or an attempted call.

clear request A control *signal* sent to a *network* by a *terminal* to terminate a *circuit* or *virtual circuit* connection.

clear request packet An instruction given to the *network* in a *packet switching system* by a *calling terminal*, when it wishes to deactivate a *virtual circuit* between it and another *terminal*. ⇨ *call request packet* and *call accepted packet*.

clearing The sequence of events associated with the disconnection of a *call* and enabling the two *terminals* concerned to return to the *ready state*.

clearing phase The action of releasing a connection in a *circuit switched* call, at the completion of the *call*, in response to a *clear request* issued by a *terminal*. Same as *call clearing*.

client layer In specifying a *data network* it is usual to provide for different levels of control as a hierarchy of procedures. An *ISO* standard, in fact, defines seven levels

35

clock

of control. ⇨ *ISO reference model for open systems architecture.*

Not all levels are provided by the *network*, but it is usual to provide at least the two lowest ISO levels: *physical link layer* and *data link layer*. Other levels may be provided by the *network operator* or be left to *users* of the network. The levels left to be provided by the users are referred to as the client layer.

clock A device which emits pulses to synchronize the operation of system elements in a digital system.

In a *synchronous digital system*, pulses are transferred at a series of discrete intervals, and devices within the system can obtain access to others by observing an exact timing sequence defined by the clock.

closed user group (CUG) A system in which *users* of a service in a *public network* can only make *calls* to, or receive calls from, predesignated *subscribers* forming the same group. Note it is possible for a user to belong to more than one CUG.

closed user group indicator Information included in a *data transmission* sequence to indicate whether the *calling party* belongs to a *closed user group.*

closed user group with outgoing access Same as *closed user group*, except that a *subscriber* in the group can call other subscribers outside the group, but may not receive *calls* from outside.

cluster A concentration of devices at a point in a *network*; e.g. two or more *data terminals* (*DTE*) connected to a *concentrator* which controls their *interface* to other terminal devices over a *communications channel.*

cluster controller An item of equipment which is sited at a remote location to concentrate a number of *terminals* at that location, and which handles the communication functions between those terminals and a *data network* or *host processor*. It

may also include certain *application programs* and sections of a *partitioned database*. The more complex cluster controllers may be programmable.

CØ A standard subset of *codes* defined for *videotext systems* to provide for various *control* functions including *cursor control, format effectors*, and *transmission control* character.

co-axial cable Sometimes used to refer to a single conducting *channel*, which more correctly should be known as a *co-axial tube*. Co-axial tubes are often grouped into cables; i.e. a number of tubes are housed in the same cable duct.

co-axial cable loop A *transmission path* in a *local network* utilizing a *co-axial tube* as a *cable loop.*

co-axial cable interface specification A specification which defines the method for connection of *nodes* or *stations* to a *co-axial cable loop* as in a *local area network*. This specification governs the *physical interface* and the correct behaviour of a station and includes electrical, mechanical and logical aspects.

co-axial tube A *transmission link* constructed of a pair of *conductors*, held in position by insulating material such that the physical distance between the conductors is maintained. The conductors are arranged so that one is a core within an outer sheath, or tube, formed by the other. The outer conductor operates as a shield to reduce the electrical interference and *crosstalk.*

For *telephone* communication, co-axial tubes are often bunched together to form a cable, and the shield then becomes ineffective at frequencies below 60 kHz. Co-axial tubes have been used up to a *bandwidth* around 60 MHz, which is the equivalent of 10,800 *telephone channels* in a *multiplexed* system.

code Any *character* or group of char-

36

acters forming a specific meaning. Any system of rules to which information must conform in order to be transmitted, received, and/or processed.

code compression A technique used in the storing and transmission of information to save storage space or transmission time across a *network*. For example, *graphic information* can be reduced for transmission by converting lines or curves into co-ordinates known as *picture description instructions*, which can be used to recreate an original picture in a *receiving terminal*.

code conversion A process whereby *data* constructed in accordance with the rules of a particular *code* is transformed into a format and structure required in another code.

code dependent system In *data communication*, a system which is dependent upon the use of a particular *data communication code* used by the *terminals* connected to the system. The system will not function correctly if other *codes* are used. Contrast with *code independent system*.

code independent system A *data communication* system which can operate correctly, irrespective of the *data communication codes* used by *terminals* connected to the system. Contrast with *code dependent system*.

code insensitive system Same as *code independent system*.

code sensitive system Same as *code dependent system*.

code-string A sequence of *bits* (*binary digits*) transmitted to or from a *terminal*, or some other element in a *transmission system*, to convey status or *control information*. This practice has arisen with the development of high-speed digital tech-

niques and enables a single connection to be used for a number of different control *signals* which otherwise would have to be conveyed by discrete *circuits*.

codec A device which incorporates encoding and decoding logic in the same assembly. ⇨ *encoder* and *decoder*.

coherent detection One method of *detection* used in *demodulation*, in which the original *carrier signal* is only partially suppressed for transmission along with a *sideband* signal. The method allows most of the energy to be concentrated in the *message signal*, but sufficient energy is used to transmit the *carrier* for creation of a *reference wave*.

collection charge The charge made by a communications authority, within its area of jurisdiction, to the *subscribers* within the area for the use of communications services which may terminate within the area of another authority.

collision A form of *contention* which arises in a *local loop* when two or more *stations* try to transmit at the same time. Due to the high *data rates* possible in such systems, collisions are usually avoided by allowing one station to defer to another without noticeable delay to *users*. Collisions can arise because at the beginning of transmission, before a *signal* has been propagated to all parts of the loop, there is a period in which no station has officially acquired the loop. This is known as the *collision window*. The resolution of collisions usually entails terminating transmission and scheduling retransmission for some randomly selected time. Fragments of *frames* which may have been received incomplete are rejected by *receiving stations*. The retransmission is repeated until successful.

collision resolution A procedure which is followed to recover from a situation in which two or more *stations* start transmitting simultaneously in a *local loop*. ⇨ *collision*.

collision window A period of time during transmission on a *local loop* in which there can be undetected *contention* for the *physical channel*. ⇨ *collision*.

colour television receiver A *television receiver* in which there are three *guns* corresponding to red, green and blue outputs from a colour television signal. The beams from the three guns are designed to scan separately over red, green and blue phosphor dots on the screen of a *cathode ray tube*. The dots are placed close together in sets of three known as *triads*. Combinations of colours are thus generated to represent the range of colours captured by the television camera. There are about 500,000 triads on a single screen.

command Any instruction issued to a *network* by a human operator, or by a logical process within an automated system, to indicate to the network that a particular *control procedure* is to be invoked.

common carrier A private or public corporation responsible for the provision of *telecommunication* services in a given territory, and who provides access to these facilities at appropriate hire charges to enable private or business *users* to communicate with one another.

A common carrier is normally not concerned with the content of the messages carried by its services, but is concerned that connections to its services are made using authorized equipment and protocols.

common carrier costs This refers to the true operational costs of *telephone* or *data networks* as experienced by the public authorities operating such facilities. The expression distinguishes the cost factors from the charges rendered by the carrier authority as *tariffs*. An economic analysis based upon costs of a service might produce a different conclusion from an analysis based upon the actual tariff.

common-channel exchange An exchange which utilizes a *common-channel signalling* system.

common channel signalling A method of providing control in a *telecommunications network* in which several hundred traffic *circuits* may be controlled by a single pair of signalling *channels* along a particular route. The route may contain several *exchanges*, each of which must be equipped with centralized control to respond to the *signalling* functions. Common channel signalling reduces the cost required for separate signalling units in each exchange, and permits greater flexibility for the future development of the service. Also known as *centralized control signalling*, and contrasted with *channel associated signalling*.

common control The *control* function in a *switched telephone exchange* is responsible for recognizing *call requests* from *users* and for managing the correct set-up, maintenance and clearing of *calls*. In early forms of switched systems, these controls were associated with signalling units which were permanently associated with each individual *line*. With the advent of *digital exchanges*, the signals concerned with control functions are handled in high-speed *digital computers* which have fast electronic systems which can work for a few milliseconds, first on one call and then on another. Thus, all the lines in an exchange may be controlled by a single common system. This can serve to reduce greatly the cost of the overall signalling system and lead to flexible and more economic maintenance procedures. This method also allows control signals passing between *exchanges* in a *network* to be handled along *channels* which are dedicated to provide *control information*; therefore, control information is not transported within the same channel as *message information*. ⇨ *common channel signalling*.

commonality Pertaining to the way in which two devices or systems can work in harmony, or to the ease with which one can replace another. ⇨ *compatibility*.

communication channel A link between two *terminals* to provide a *go* or *return path*. ⇨ *channel*.

communication interface A specification which defines the necessary conditions for connecting two parts of a system, or two separate systems, which have different functions. This requires a definition of:

The *logical interface* (e.g. what various *signals* mean and how they represent functions);
The *mechanical interface* (how physical components connect, such as the significance of pin positions on a plug connector);
The *electrical interface* (the strength, frequency and duration of signals across the *interface*).

A considerable amount of work has been done by the *CCITT* and other bodies to standardize interfaces so that systems, developed in different countries and by different manufacturers, can communicate. This work includes the publication of specifications for *interchange circuits* and *interchange specifications*.

In most forms of *telecommunication* over *data networks*, the *data terminal* (*DTE*) is interfaced to the *line* by a *data circuit terminating equipment* (*DCE*), and it is necessary to consider the interface between DCEs at each end of the *channel*, and between each DCE and its DTE. An example of such a specification is given by the CCITT *V24* recommendation which is a *hardware* interface for establishing a *call* and ensuring appropriate *error detection* and *correction* takes place. This is sometimes known as the *physical level interface*.

person/machine interface: this defines the way in which a human *user* interacts with a system (e.g. the rules governing the operation of the *keyboard* of a *teletypewriter*).

software interface: a high-level specification which defines information flows and application procedures.

Efforts are continually being made to improve standardization of these different levels of interface, and an example of this work is described under the heading *ISO reference model for open systems architecture*.

communication interface standard A standard *protocol* designed to ensure that *terminals* using the same logical, mechanical and electrical connections can communicate with one another and with *public data networks* (*PDN*). Examples given in this book include *V series* interface and *X series* interface.

communications control Relating to the co-ordination and management of any communication *network*, and the *signalling systems* used to set up, maintain and clear *calls*; as well as the *flow control* of *messages* in networks which handle *data* or *text*.

communications controller A device used in a *data network* to manage all *data link control* activities between a *main frame* (or *host processor*) and a large distributed *terminal* population. Usually it would be sited adjacent to a host processor to which it would be connected by a high-speed link. Some communications controllers take responsibility for *routing*, *dialogue management* and *protocol conversion* also, and remove as much of the communications load as possible from the host processor.

Communications controllers are really programmable *computers*, having a great deal of power and flexibility to be applied in a number of different networking environments. Also known as *communications processor* or *front-end processor*.

communications processor In a *data processing system*, a *computer* which *interfaces* the main *host processor* to a communications *network* to handle all the *traffic* to and from the host processor, and to communicate with *terminals* or other processors in the network. A communications processor will not normally process *data* (i.e. look at the *application* content of *messages*), but is reserved specifically to managing the *telecommunication* functions related to its host. Also known as *communications controller*.

communications satellite A *station* placed in orbit around the earth to provide *transmission channels* for *traffic* to be transmitted over great distances; e.g. intercontinental traffic. The *satellite station* is equipped with *antennae* which are able to receive radio beams from an *earth station* transmitter, and to retransmit the *signals* to another earth station. The orbit of the satellite is chosen so that signals can be propagated between specific locations.

A worldwide system of satellites has been created, and it is possible to transmit signals around the globe by bouncing them from one satellite to an earth station, and thence to another satellite.

Originally designed to carry *voice traffic*, they are able to carry hundreds or thousands of separate simultaneous *calls*. The *radio signals* used to carry the calls contain great numbers of channels by *multiplexing* techniques.

These systems are being increasingly adopted to provide for business communications, including the transmission of traffic for *voice*, *facsimile*, *data* and *vision*.

So far, satellite signals have been distributed through the normal *earth networks* when received at the *receiving earth station*.

In the USA, business organizations are hiring facilities to make use of the systems provided by satellite operating companies. The companies wishing to receive signals can do so directly from a satellite station by using a small dish-shaped *aerial* placed upon the roof of their office building.

One of the first commercial satellites (Early Bird) was launched in 1965 and it carried 240 separate channels.

In the early 1980s, there are approximately 70 separate satellite stations in earth orbit. For example, the Intelsat network has 12 satellites and covers more than 150 separate countries. The more recent satellites each carry approximately 13,000 separate channels of communication.

The European Space Agency has plans to launch satellites with more than 240,000 separate *telephone channels* or, alternatively, five *television channels*.

The television channels would be able to provide direct broadcasting to homes equipped with special dish aerials placed on the roof. This will enable TV broadcast authorities to provide programmes far beyond their present geographic boundaries.

It is probable that future satellite services will enable a great variety of information services to transmit directly into the home, possibly including personalized *electronic mail*.

compatibility Pertaining to the degree of *interworking* possible between two devices or systems. If an element in a system is fully compatible with the functional and physical characteristics of a system, it can be placed into the system without any effect upon the grade of service experienced by *users*. ⇨ *transparency*.

compiler A *software* system which converts statements written in a particular *programming language* into machine code instructions which can be recognized by a particular type of *computer*.

composite video signal A signal which includes video information, plus the *synchronization pulses* used to control the positioning of the *raster* in a *TV receiver* or monitor used to play back a video *message*. Also known as *visual message signal*. ⇨ *video signal*.

composition coding A technique used for the display of *character sets* for languages which encompass diacritical marks, such as accents in Latin-based language forms. The technique requires the coded representations of the *character* and the associated diacritical mark to be transmitted separately and composed together to form the desired composite character in the display terminal. This is a method recommended for use in *videotex systems* where the full range of character sets used in a particular country must be displayed. Compare with *dynamically redefinable character sets*.

computer Any machine which can accept *data* in a certain form and process the data to supply results or to control a process. Generally, an electronic device in which inputs and outputs may be in *digital* or *analog* form, and in which the program controlling the operation of the *computer* may itself be modified by logical actions taken under program control.

A program for a *digital computer* is a set of instructions which can be interpreted by the central processing unit in the computer, and which results in data being manipulated as required to produce a desired result. A typical computer will respond to a hundred or more basic types of commands such as ADD, SUB-TRACT, COMPARE. Each of these commands may refer to items of data in specified storage locations. The *users* who wish to make the computer perform a specific task have to write their programs as a series of step-by-step instructions. In practice, few programmers write their programs directly in the language that the computer understands. Instead, various levels of other programs, known as system software, are available which act as translators between the machine and the end user's program. Computers perform instructions in millionths of a second, and a complete program can be performed in a very short time. Because a computer can store very large amounts of information, and because it can, under program control, modify its action to react to changes in external events, it has become a predominant tool in the modern world, and of particular influence in the increasing efficiency and functioning of modern communications systems.

computer aided design system A *computer* system which has been specially designed to assist industrial designers in their work, allowing designs to be constructed and visualized on display screens. Such systems are used in diverse industries, including automobile manufacturing, aircraft design, town planning and civil engineering.

computer conferencing A system in which a computer is used as a central coordinating and storage device to manage a conference about a specific subject. The conference is usually available only to members of a defined group, who are given the right to present propositions and arguments as messages, which will be stored and made available in context and sequence to other members of the group.

In this form of teleconferencing it is not necessary for all members of the group to communicate with the computer at the same instance. Members can participate in the conference as and when they are able and free from other activities. ⇨ *teleconferencing*.

computer networking The use of *networks* to link *computers* together so that they can share a workload, or allow *users* connected via *terminals* to a particular computer to have access to facilities and services provided by other computers in the network.

concentrator
1. A device which enables a number of *calls* arising on individual *lines* to be concentrated on to a single line using *multiplexing* techniques.
2. An item of equipment in a *telephone exchange*, which receives calls arising on *local lines* in a particular area, and directs the calls to a *distributor* for transmission to *outgoing trunks* or outgoing *local lines*.

In a *circuit switching* system it has to be possible to switch every *incoming line* of an exchange to every *outgoing line*. For an exchange with 1000 incoming and outgoing lines, this would entail a matrix of 1,000,000 *crosspoints* at which switching could take place. However, more efficient internal *architectures* are possible which reduce the number of crosspoints whilst maintaining a low probability of *call block-*

ing. A typical design might reduce the number of crosspoints in the example above to 30,000. This improvement in efficiency can be achieved by having three stages of switching: (a) *concentrator*, (b) *distributor*, (c) *expandor*.

The first stage concentrates the incoming lines on to a smaller number of internal links before passing calls to the distributor.

The second stage switches calls and then passes them to an expandor, or to an outgoing *trunk route*.

The expandor accepts the switched calls from the distributor, and connects them to the appropriate outgoing lines in the *local network*. ⇨ *grouping*.

conductor An element used to carry an electrical current; e.g. a pair of copper wires used to carry an electrical *message signal* from a *terminal* to an *exchange*.

configuration control A set of functions performed at a *network management centre* to control the availability of paths and resources in the overall *network* in an orderly fashion. For example, some new *transmission paths* and *nodes* may be attached to the network, and some may be temporarily withdrawn for maintenance purposes. A set of complex facilities provided by *hardware* and *software* are usually available to a human operator to put into effect configuration control procedures.

confusion signal A *signal* passed on a *backward channel* to indicate that a *node* in a *network* is unable to act upon a signal received because the signal is not a reasonable request.

congested system If a system has insufficient capacity to carry the *traffic* required in a given period of time, it is said to be congested.

connect charge A charge made to a *user* as part of the *tariff* for utilization of a service, in which the user pays a fee for each minute (or part of a minute) that he is connected to a *port* in the system.

to a *data terminal* to advise the *terminal* that the connection is about to be made and that the *ready-for-data signal* will follow.

contention A situation which arises when two *message signals* attempt to use the same physical resources; e.g. a *time slot* or a physical *circuit* at the same instant. In some systems it may be a deliberate policy to allow contention, and a formal procedure may be established to allocate the *line* temporarily to *terminals* on a first-entry basis. ⇨ *collision*, and contrast with *polling*.

contention resolution A procedure which is observed automatically by a system to deal with situations in which two *messages* attempt to use the same physical resources or *time slots* simultaneously.

continental circuit Any *international circuit* between two *exchanges* in different countries but in the same continent.

continuity check A check made upon a *circuit*, or series of circuits, to verify that a particular *transmission path* is not broken.

continuity-failure signal A *signal* which is passed in the *backward direction*, indicating that a *call* cannot be completed because a continuity check has revealed an interrupted *circuit*.

continuous receiver Any device, such as a *teleprinter*, which can record a series of *messages* received line by line, without the need for operator intervention between messages.

control This is a role performed by an *exchange* or a *node* in a communications *network*, and is concerned with the functions which govern the orderly selection and maintenance of paths through the *network* to allow a *call* to take place. These functions include:

recognizing a *call request* from a *terminal* receiving the *address information*

connect signal A *signal* transmitted in the *forward direction* at the beginning of a *call* to secure a *circuit* to switch the call.

connection charge A single payment charge made by a new *subscriber* for being connected to a communications service provided by an operator.

connection in progress In a *data network*, a *control* signal from a *modern* (or *DCE*) translating the *address* to identify a route selection of a specific path
sending *control* signals to the *line*
monitoring the *call answered signals*
clearing the connection

In the interests of efficiency, control equipment is usually shared over several lines.

control character Any *character* occurring as a *signal* in a context which causes a procedure or operation to be stopped, started or modified.

control circuit Any *circuit* used to convey supervisory information to co-ordinate transmission taking place on another circuit. For example, in transmitting television pictures from an outside broadcast event to a studio, one or more control circuits may be used.

control information *Data* or *signals* carried in a *network* to support the *control* functions.

control input A *signal* provided by the action of a *user*, *subscriber* or operator in a communications system, and having significance in respect of the recognition of a *call request*, or the set-up, maintenance and termination of a *call*.

control procedure In any communications service, a method by which *signals* are sent in a *forward* or *backward direction* in accordance with a predetermined order to ensure co-ordination between *users* and the *network*.

controlled maintenance A systematic method of maintenance using sampling and analysis techniques to reduce *corrective maintenance* and improve efficiency in *preventive maintenance*.

controller A device at the heart of a communications system which is responsible for co-ordination of the system by receiving, interpreting and transmitting *signals*; e.g. responsible for switching *calls* in a controlled manner.

controlling exchange A designation given by telephone administrations when participating in international *calls*. The *exchange* which sets up calls and decides the sequence in which they are connected is the controlling exchange, and usually it is the *international exchange* to which a *calling party* is connected.

conversation The process which takes place when two *data terminals* exchange information; a period of activity which is preceded by the *call establishment* process, and followed by *call clearing*.

conversation state The *state* that exists in a *telephone network* when a *called terminal* has responded to a *calling signal*, and a *circuit* is made to enable a conversation to take place.

conversation time In the operation of a public *telephone system*, the time interval between the *answer signal* and *clear forward signal* being recorded, at the point where the recording of the *call duration* takes place.

conversational mode A method of communication involving two-way communication between two *terminals*, in which both *users* may act as a *message source* and a *message destination*. The dialogue is carried out in accordance with a *protocol* which serves to ensure that the terminals are co-ordinated and give appropriate responses to one another to secure the safe and intelligible transmission of *messages*.

CORAL 66 A general-purpose *programming language* (based upon Algol 60, an earlier language), and designed for the UK Ministry of Defence for use in military command and control systems.

corporate networks A *telecommunications* system installed within an organization to meet its needs for communication between staff internally, or external organizations with whom it transacts business on a regular day-to-day basis. May make extensive use of public facilities and privately owned facilities.

correct bit A *bit* (*binary digit*) of information correctly received over a *circuit*, as against an *error bit* in which the significance of the bit has been reversed; i.e. a 0 becomes a 1 or a 1 becomes a 0. ⇨ *transmission errors*.

corrective maintenance Work carried out to repair any device which has failed, such that the device is considered to be unsuitable for maintaining the operation of the system to the level of quality required.

counter A device which records a number of events to assist in the control and co-ordination of a logical process. For example, a device which records the number of *blocks* of *data* sent to a distant *terminal*.

country code A *code* used to prefix a telephone number and to designate the country in which that particular *subscriber* is situated. The code is used to facilitate *international trunk dialling*.

CR ⇨ *carriage return*.

cradle switch The contacts which are activated when a telephone *handset* is raised or put down. Sometimes called a *switch hook*.

CRC Abbreviation of *cyclic redundancy check*.

crossbar exchange An *exchange* in which

a bar rotated by a *solenoid* is used to make switched connections between *circuits* to connect *incoming calls* to *outgoing lines*. Typically, a single *crossbar switch* provides a matrix of 10 by 20 individual switches, which are contact points activated by a complex arrangement of levers. The selection of a particular path is achieved by rotating the bar. Based upon the operation of electromechanical devices, these exchanges are being gradually superseded by *electronic exchanges* which are faster in operation and easier to maintain. ⇨ *switching equipment*.

crossbar switches An electromechanical device used in *telephone exchanges* which carry *voice analog signals*. It consists of contacts arranged in a matrix, which are activated by a metal bar which is, in turn, rotated by the action of a *solenoid*. These devices are constructed in large matrixes to connect *incoming lines* and *outgoing lines* to establish *calls* in a *circuit switched exchange*. These are being replaced in the present generation of exchanges by *electronic exchanges* which are more efficient used in conjunction with digital forms of transmission.

cross-office check A *continuity check* made through an *exchange* to verify that a *transmission path* exists.

crosspoint A switch forming part of a matrix to switch *circuits* in an *exchange*. Originally, crosspoints were electromechanical switches, but have progressed through to *reed-relays* and now to *electronic crosspoints* with the advance of switching technology.

crosstalk An undesirable condition in which *message signals* from one *channel* are overlaid upon another, physically adjacent channel. This condition is caused by stray electric and magnetic *fields* which are generated when *AC signals* are transmitted along a *line*. ⇨ *transmit-to-receive crosstalk*.

crosstalk, intelligible *Crosstalk* which results in intelligible *signals* being transferred from one *circuit* to intrude on another.

crosstalk, possible *Crosstalk* components which exist but do not intrude on the *user* at the point at which they have been measured, but may intrude at another point.

crosstalk, unintelligible *Crosstalk* which results in intrusive unintelligible speech components being transferred from one *circuit* to another.

CUG Abbreviation of *closed user group*.

cursor control In any system which uses *visual display* units for the presentation of information, control *codes* are used to govern the positioning of *characters* and to control the format of documents presented on display screens. The position for the next character to be displayed is always indicated by the movement of a character known as the cursor, and formats are effected by cursor control instructions.

customer reference number A reference number which is unique to each *user* of a service, and which enables *usage charges* and utilization statistics to be attributed to individual users.

customer's loop Synonymous with *subscriber's line*, or *local line*.

cyclic code A coding system used for *error detection* in which a calculation is performed upon each *block* of *data* to be transmitted and a remainder is derived to be appended to the data as a *check character*. The same calculation is performed at the *receiving station* and a comparison is made to see that *bits* have not been lost or inverted during transmission. The method relies upon treating the bits of the data as a pure *binary number*, and the remainder is produced by dividing by a *generating polynomial*. A *CCITT* recommendation (*V41*) advocates a 16-bit polynomial of the form $x^{16} + x^{12} + x^5 + 1$. ⇨ *cyclic redundancy check*.

cyclic redundancy check (CRC) A procedure used in checking the accuracy of information *frames* transmitted over a *data link*, in which a series of *bits* known as the *frame check sequence* (*FCS*) is derived and appended to each frame. The FCS is computed as a function of all the *fields* contained in the particular frame, prior to transmission. A *receiving station* then checks the accuracy of the frame by attempting to derive the same FCS from the received information. If the procedure fails to validate the FCS field, a retransmission of the frame is requested.

D

data Any values, *numbers, characters* or symbols which have been arranged to represent information in accordance with predefined rules. It is generally noticed that the word 'data' is used often as a singular noun.

data assurance The procedures in a communication system which are concerned with the detection and correction of *transmission errors*.

data channel A *transmission channel* used to carry *data* to provide a means of communication between two points.

A data channel may be an *analog data channel*; i.e. providing a path for alternating current *signals* which are converted at each end of the *circuit* into *digital signals* for a *terminal* by a *modulator/demodulator* (*modem*).

A data channel may alternatively be a *digital data channel*; i.e. providing a path for signals made up of discrete pulses, with a *network interface unit* at each end to *interface* to the terminal.

In strict definition, a *channel* provides a means of one-way transmission and two channels are required to form a means of two-way communication known as a *data circuit*.

data circuit A two-way means of transmission, consisting of two *channels*, allowing for the transfer of *data* between two *terminals*. A data circuit may carry *digital* or *analog signals*. In the former case, the *data terminal* (*DTE*) at each end of the *circuit* is interfaced to it by a *network interface unit*. In the latter case, the terminal (DTE) is interfaced by a *modem*. ⇨ *data channel*.

data circuit terminating equipment (DCE) The equipment installed to *interface* a user's *data terminal equipment* (*DTE*) to

a communications line. It is not necessarily a separate freestanding item of equipment. The DCE provides for the establishment, maintenance and termination of a call and also the signal conversion necessary between the DTE and the *line*.

A *modem* is an example of a DCE but, when connecting a terminal to a specialized *data network*, a device known as a *network terminating unit* or *network interface unit* is used. In general, modems are used for connection to telephone networks where *analog signals* are used in transmission, and a network interface unit to a *public data network* (PDN) where digital signals are used.

data collection The process entailed in collecting information about events and *encoding* them for transmission and reception, prior to subsequent processing to achieve an end objective of an information system.

data communication The whole range of practice concerned with the transmission of information which has been encoded specifically for the purpose of transmission (e.g. in the form of a *data communication code*), and including the encoding, transmission, *routing*, monitoring, checking and correction of errors in the process of transmitting and receiving *data*.

data communication code A *code* used to represent *characters* of information as groups of *binary digits* (*bits*) and containing a system of *binary notation* to represent the 26 characters of the alphabet, numerals 0 to 9, and a range of special symbols and punctuation characters. There are a variety of codes in use, but there is a movement towards international standardization of such codes; e.g. ⇨ *International Alphabet Code no. 5*. The codes are designed for transmission of digital information for telegraphic or *data transmission* purposes, and also include a number of codes to perform special functions, including: *transmission control, format effectors, information separators, device control.*

Most systems use 6 bits to represent each character, but some systems use 8-bit codes. Most data communication codes have been developed from the *telegraph codes* which preceded the evolution of computer-based systems.

data communication system Any system for communication in which information is transmitted between locations and, for the purpose of transmission, is represented in coded form as *electrical signals.*

data compression A technique used to improve the efficiency of transmission by reducing the number of information *bits* which need to be transmitted. For example, in some graphic display systems, arcs and lines can be transmitted as co-ordinates which are expanded in the *receiving terminal* to recreate images stored at the *transmitting terminal.*

data connection The process entailed in switching to connect together a number of *data circuits* to provide a path for *data transmission.* The term also refers to the set of physical resources used in making a connection between two *data terminals.*

data element A *field* in an information record, or part of a *message signal*, having a logical relationship to other items of information; e.g. account number, customer name, price, quantity ordered.

data encapsulation The functions in a communications system concerned with control of *data* in a *transmission link*; e.g. establishing the boundary of *frames*, handling the generation and recognition of source and *destination addresses*, and the detection of *transmission errors* in the *physical channel.*

data encoding Any process by which *data* is converted from one form, to be repre-

sented in another form, for the purpose of transmission.

data entry terminal A terminal specifically designed to enable *data* to be collected and prepared for transmission over a *circuit.* It is usually equipped with a *keyboard*, a printer or display screen, and a *modem* or *network interface unit* to enable it to send and receive information over a *network.*

data flow control (DFC) An element in a *packet switching network* which has responsibility for regulating the direction and flow of *packets* by end *users.* It manages the relationship between the users to ensure that an appropriate *conversation* (pattern of intermittency) takes place, and maintains the relationships of *messages* to the separate *packets* of which messages are comprised.

data link The set of physical resources which are connected together to form a path for communication of *data* including the *data terminals* and all interconnecting resources.

data link control (DLC) A device and associated *protocol* which ensures that a *data transmission* is free of errors. In its simplest form it examines groups of *bits* arriving at a *receiving terminal* and checks against a predetermined DLC protocol to see whether the bits in the group have been misplaced or dropped. A group of bits is referred to as a *frame.* If a frame is found to be in error, a retransmission is automatically requested.

As each frame is initially transmitted, bits are added to form a certain parity in the *bit stream* in accordance with the DLC protocol, and the receiving DLC checks against this protocol. ⇨ *high level data link control.*

data link controller That part of a communications device which is responsible for *data link control*; i.e. for ensuring error-free *data transmission.*

data link escape (DLE) An international *transmission control code* which changes the significance of other *characters* which follow after it in a transmission sequence.

data link layer In a *network architecture* this level of control includes functions which are independent of the physical medium used for the *channel*, but define the basic procedures for *data encapsulation* and *link management*. This corresponds to the second level in the *ISO model of architecture for open systems interconnection*.

data network A communication system used for *digital data transmission* and which may use *private networks* or *public data networks*, but having the potential to provide multiple *access paths* between *users*.

data phase That period of time during a *call* in which *data* may be exchanged between *data terminals* interconnected over a *network*; i.e. excluding time for *call establishment* and *call disestablishment*.

data printer Any form of *terminal* used for printing a *hard copy* of information received over a *communications network*.

data processing system Any system designed to perform operations upon *data* by some form of automatic processing and control. Designed to produce an ordered result from raw data.

data processing terminal A device used to transmit or receive *data* over a *network* usually in the form of *digital signals*.

data rate A term which relates to the speed at which *circuits* or devices operate when handling digital information. For example, a particular *transmission channel* might be rated at 2400 *bits* per second, where a bit is a pulse representing a *binary digit*. ⇨ *modulation rate, data signalling rate* and *data transfer rate*.

data security Procedures established to protect a system and its *users* against the

intentional or unintentional misuse of *data*; e.g. disclosure of confidential details or the modification or destruction of information.

data segment The part of a *packet* or any *message signal* which contains *data* rather than *address* or *control information*.

data service unit A simplified *modem* for transmission of *digital data* over a *private line*, or for limited distance communication in which it is not necessary to comply with all the requirements for a *high speed modem* using the *public switched telephone network*. Other names given to such devices include *baseband modem, line driver, line adaptor* and *limited distance modem*.

data set ready A *signal* sent by a *modem* or *network interface unit* (*DCE*) to the *line* to inform a *calling terminal* that the DCE has received a *data terminal ready* signal from its associated *data terminal* (*DTE*).

data signal Any *signal* which consists of *data* arranged to represent information, and which may include *check digits* added to the data to provide *error control* facilities.

data signalling rate A term used to express the rate at which information can be transmitted over a *circuit*. In most digital systems, the information is transmitted as a series of pulses having the significance of *binary digits* 0 or 1, according to the polarity of each pulse. In such a case, the *information rate* and the *data signalling rate* are the same.

In some systems, the *amplitude* of pulses may be varied to represent different values, even though the duration of pulses remains constant. Thus, for example, a pulse may have the value of 00, 11, 01, or 10 according to amplitude. This is an example of *multistate signalling*. If the duration of each pulse is 20 milliseconds, the modulation rate is 1×0.02 *bits* per second; i.e. 50 *bauds*. But the data signalling rate in this example is 100 bits per second. ⇨ *data transfer rate.*

data signalling rate transparency Refers to the capability of a *network* to provide compatibility between *terminals* operating at different *data rates.*

data sink A term referring to a device (e.g. a *paper tape punch* or a line printer) which receives information over a *circuit* from a *data source.*

data source Any device which generates *data signals* to be transmitted as information over a *circuit* or *network*. Compare with *data sink.*

data switching exchange A set of equipment designed to make connections to switch *data traffic* from one *terminal* to another. It may be a *circuit switching* system and/or a *packet switching system.*

data terminal Any device capable of sending and/or receiving information over a communications *network*. Generally speaking, a device capable of sending or receiving digital information; but *analog* devices, used to measure phenomena remotely, also can be classed as data terminals. A data terminal can range from a simple terminal to a very complex *computer*. Same meaning as *data terminal equipment (DTE).*

data terminal equipment (DTE) A *terminal* or *computer* attached to a *data network* as an end *user node*. Such a device must operate in accordance with the defined *protocol* for the *network*. For example, under *X25 network architectures*, it must *interface* with *packet level protocols* which govern the size, sequence and format of *packets*, and *frame level protocols* which manage error-free transmission of packets to and from the network.

The DTE is responsible for these *high level functions* rather than *physical level* functions which are performed by the *data circuit terminating equipment (DCE)*. The

DCE is a *modem* or *network interface unit* which connects the DTE to the network.

data terminal ready An *interchange signal* between a *data terminal (DTE)* and its *DCE* which signifies that the DTE is operable and ready to receive *data*. ⇨ *calling indicator signal.*

data traffic *Message signals* which represent information pertaining to a *data processing system* rather than *speech signals*; and implying the use of *digital transmission.*

data transfer The process of transferring information from one location to another in a communication system, and often cited to distinguish between other phases in a *call* sequence, including *call establishment* and *call clearing.*

data transfer phase The period in which a *user's data* is transferred between two *terminals* in making a *call* over a *switched circuit* network. It is distinguished from the *call establishment* phase in which an *access path* is created from the *calling* to the *called terminal*, and *call clearing* phase in which the *circuit* is released.

data transfer rate A term used to express the rate at which information is received over a *circuit*, excluding such signal elements as are used to synchronize transmission; e.g. excluding *start bits* or *stop bits* used in telegraphic transmission to denote the beginning and end of characters. The data transfer rate is thus less than the *data signalling rate* possible over the circuit. Another factor to be discounted in arriving at the data transfer rate is the number of redundant *characters* and extra transmission sequences required as the result of the *error control* procedures operated over the circuit.

The term data transfer rate is used as an accurate statement of the performance of a circuit in transferring information, and is related to a particular method of transmission under given conditions. The rate

49

may be expressed in *bits*, *characters*, *words* or *blocks* received per unit of time. Also known as *information rate*.

data transfer requested signal A *control* signal from a *data circuit terminating equipment* (*DCE*) to its associated *data terminal* (*DTE*), indicating that a distant *terminal* has requested to transfer *data*.

data transmission The process or techniques concerned with transmitting information as *digital pulses*; i.e. the movement of information in coded form as electrical energy. ⇨ *digital data transmission*.

database In *data processing*, a *file* (or files) organized in such a way that a variety of *users* can update or enquire of the file for different purposes, using *computer* procedures, which to the user appear to be independent of the file structure. A file which is not designed to satisfy a specific limited application.

database management Relating to a *computer* system operating to control the recording, analysis, indexing, storage and retrieval of *data*. Implying a method of *file* organization which allows for the efficient production of required results in response to both standard and ad hoc requests.

datagram In a *packet switching system*, the simplest form of transmission which can occur is the transmission of a single *packet* from one *user* to another, requiring no response. It is sometimes referred to as a datagram.

Datex The name given by the Deutsche Bundespost to a range of *public data transmission services* available to *subscribers* in West Germany. The specific service is denoted by a suffix; e.g. Datex – P represents the *packet switching service*.

dB Abbreviation of *decibels*.

DC ⇨ *direct current*.

DC1, DC2, DC3 and DC4 ⇨ *device control codes*.

DCE Same as *data circuit terminating equipment*, it corresponds functionally to a *modem* or *network interface unit*, and its purpose is to handle all *control* activities entailed in connecting a *data terminal* (*DTE*) to a network.

decentralized control signalling A system of *control* in which control *signals* related to *data transmission* on a particular *circuit* must be carried on that circuit.

Also known as *channel associated signalling*. Contrast with *centralized control signalling*.

decibels A unit of measure for the power of a *signal*, or a measure of the *attenuation* produced upon a signal in any part of a communication system.

It is usual to relate power levels to a standard point, and these points vary and are defined by the particular system or particular measuring device being used.

As an example, a reference of one milliwatt may be taken and power levels would be described as P dBm, where P is positive for power in excess of a milliwatt, and negative for less.

decimal digit A *digit* in decimal notation; i.e. from the set of digits 0, 1, 2, 3, 4, 5, 6, 7, 8 and 9.

decision feedback system A system based on the *error control* principles of *ARQ* or *automatic repeat request*.

decoder

1. A device which interprets information represented in a defined *code* and generates output into a form required for another processing operation. For example, the reconstruction of samples from *character signals* in *pulse code modulation*. Also a device which converts *electrical signals* received at a *receiving station* into the form

required by the *data link control* in the *receiving terminal*.

2. Hard-working lexicographer.

deference A procedure by which a *data link controller* delays its transmission to a *channel* to avoid *contention* with other transmissions using the same channel. This technique is sometimes used in *local area networks*.

DEL ⇨ *delete*.

delay distortion A form of signal impairment which arises because of a variation in the propagation time for different frequencies in a *circuit*.

delay equalizer A device used to overcome distortion arising from the differential effect of a *telephone channel* upon the propagation of *data signals* in different parts of the *bandwidth*. The delay equalizer delays the more advanced frequencies to coincide with delayed frequencies; it therefore introduces an overall delay of a few microseconds which is not significant.

delay time The *answering time*; i.e. the interval that elapses between the completion of a *calling signal* and the response by the operator or automatic equipment at the *called location*.

delayed delivery A facility which is available in certain *data networks* to allow *data* to be stored temporarily until a particular *destination terminal* is available.

delete (DEL) A special function used in a *data communications code* to allow *users* to overwrite an erroneous *character*. For example, in connection with information punched in *paper tape*, a DEL is represented by a complete row of holes in every position of a character.

delivery confirmation A notification passed to a *data terminal* which has used a *delayed delivery* facility, confirming that the *network* has now delivered the *message* to the *destination terminal*.

democratic network A *network* in which each *node* or *exchange* has *clocks* of equal status; i.e. no one clock has control over the whole network and the clock rate is defined as the mean of the clocks involved. Contrast with *despotic network*.

demodulation The process by means of which a *message signal* is extracted from a *carrier signal*, which has been used to transmit the message signal over a *network*, or as a *radio transmission*. ⇨ *modulation*.

demodulator

1. A device which receives *data signals* in *analog* form and converts them into the form of *digital signals* representing *binary digits* suitable for processing in a *data terminal equipment*.

2. A device which analyses a *signal* consisting of a *carrier wave* modulated by a *message signal*, and reconstructs the original message signal for further processing at a *receiving station*. ⇨ *modulation*.

demultiplexed A *message signal* is said to be demultiplexed when it has been separated from a *carrier signal* with which it had originally been combined for efficient transmission. ⇨ *multiplexor, frequency division multiplexing*, and *time division multiplexing*.

demultiplexing The process of separating *message signals* which have been combined for the purpose of transmission on to the same physical *transmission path* using the techniques of *time division multiplexing* (*TDM*) or *frequency division multiplexing* (*FDM*).

demultiplexor A device used to separate individual *message signals* which have been combined on to the same *transmission channel* by using the techniques of *frequency division multiplexing* (*FDM*) or *time division multiplexing* (*TDM*).

demux Abbreviation of *demultiplexor*.

de-packetizing The process which arises in receiving and handling an element of a

51

message signal over a *packet switching network*. The element is stripped of *codes* which have been added to it to assist in transmission, and is combined in correct sequence with other elements of the same *message* which have been transmitted as separate *packets*.

deserializer A device which accepts a series of pulses one after another and groups them into required patterns according to the coding structure required for output. For example, it receives *bits* in *serial mode* and outputs groups of bits in *parallel mode* to represent each *character code*.

despotic network A *network* in which the timing of all operations is governed by one master *clock*. Also known as *synchronized network* and contrasted with *democratic network*.

destination address The *station* designated to receive a specific *message signal*; or that part of a message signal which defines the *address* of the *called location*.

destination address field A *field* of information within a *frame*, which identifies the *address* of the *station* to which that particular *packet* or frame of *data* is routed.

destination terminal The *terminal* designated to receive a specific *message signal*. Same as *called terminal*.

detection The process used in a *demodulator* to reconstruct an original *message signal* from a *modulated carrier wave*. ⇨ *envelope detection, synchronous detection* and *coherent detection*. ⇨ *modulation*.

device control Processes or *signals* concerned with the physical activation of remote equipment rather than the information being carried in a communications system.

device control codes (DC1, DC2, DC3 and DC4) Special *codes* occurring within the *character set* of standard *data communica-*

tion codes and used to represent instructions to activate certain specified functions on a *terminal* device (e.g. to switch it on or off).

DFC Abbreviation of *data flow control*.

dial tone The *proceed-to-send signal* observed by the *user* of a *telephone* when he is connected to an available *line*.

dial-up connection A connection made via a switching operation in an *exchange*, by a *subscriber* who is able to dial a connection to a required destination. Any connection made by automatic means at the time required, rather than one effected by a permanent point-to-point circuit.

diallable symbol Any symbol (numeral, letter or special *character*) which can be selected by an operator using a dial or push-button *keypad* on a *telephone* instrument.

dialogue A process consisting of a series of *signals* passing between two *terminals* which are in contact with each other over a *network*, in order to perform a two-way sequence of communication. The dialogue proceeds according to a defined *protocol* which is designed to ensure an orderly exchange of information and the correct initiation and termination of the *call*.

dialogue management Once an *access path* has been established between two *users* in a *packet switching network*, a *dialogue* between the two users may take place. The access path has to be retained, by a *virtual circuit*, to suit the *pattern of intermittency* of the *call*.

In the simplest case, a single *packet* of information may travel from one user to another; or there may be a *conversation* allowing a *transaction* to be progressed, in which packets transmitted from one user result in several packets in response from the other.

Also, a so-called *session* may take place

in which a number of related transactions are progressed.

Such a dialogue has to be managed under control of *software* at each *station*. While the dialogue is taking place, it is necessary to associate related packets being transmitted and received, and to control the dialogue so that the stations send and receive packets in a co-ordinated manner as though it were a rational exchange between two people.

dibit Used to refer to two *bits* of information.

differential echo suppressor A device which suppresses reflected *signals* in a long-distance telephone *circuit* by monitoring the difference in levels between signals on the two *speech channels* of a *4-wire circuit*.

differential modulation ⇨ *differential phase modulation*.

differential phase modulation A method of *phase modulation* in which no particular value is attributed to *phase inversion* in a certain direction. A change in *phase* simply indicates a change from a previous value to another; e.g. a change from 0 to 1 or from 1 to 0.

digit A numeric element selected from a finite set of elements and in *data transmission* represented by a pulse (or group of pulses) of a certain *amplitude*, timing, duration or *phase*. The term should be used to indicate in context the radix of notation; e.g. a *decimal digit*, or a *binary digit*.

digit position The space or *time slot* into which a particular *digit* is positioned.

digit pulse A pulse corresponding to a specific *digit position* in a transmission sequence representing a *binary number*, or corresponding to a specific *time slot*.

digit rate The number of *digits* transferred in a given time interval. Care is to be exercised in the use of this term to

define the radix of notation; e.g. *binary digit*. ⇨ *modulation rate, data signalling rate* and *data transfer rate*.

digit signal In *telephony*, a *signal*, generated by dialling, and representing a specific *digit* forming part of a series giving the *address* of a *called terminal*.

digit time slot The *time slot* allocated to a particular *digit position*.

digital computer A *computer* which operates using a stored program of instructions, in which information to be processed and the instructions are represented by *digital pulses*. The information is usually recorded in *binary notation*. Such devices range in size from small *micro-computers* through to large *mainframe computers*, and can handle millions of operations per second.

digital connection A *digital path* between two *terminals* operating at a specified *bit rate* and using a switched connection through a *digital switch*.

digital data channel A *channel* used to carry information recorded as discrete pulses representing *binary digits*, rather than used to carry *analog signals*. It is possible today to convert *voice signals* into digital form by using *pulse code modulation* techniques.

digital data link A set of resources used for the transmission of information recorded in digital form at a specified *bit rate* between two locations. ⇨ *digital data transmission*.

digital data transmission Digital transmission is used for the transmission of *data* (e.g. information recorded in a form recognized by *computers*), and *digital techniques* have been designed to provide for very low *error rates*, which are not available using *analog techniques* over a normal *telephone channel*. A *digital signal* deriving from a computer, or a *terminal* (*DTE*) *digital error*

which is compatible with a computer, will produce a *signal* at discrete voltage levels as a series of pulses, e.g. at two levels, representing the 1 or 0 condition associated with *binary numbers*.

To transmit digital signals, it was common, prior to 1960, to convert this simple digital signal into an *analog signal* by means of a *modem*. The signal could then be transmitted by *frequency division multiplexing* as though it were a *voice signal* to be converted finally back to digital form by a modem at the *receiving terminal*.

At a later stage, *time division multiplexing* techniques were introduced on *trunk routes* to enable signals to be transmitted in digital form, and *network interface units* were used to *interface* each terminal to the digital *network*, thus allowing digital transmission throughout the network.

These techniques enable a simple *voice channel* to support more than 20 digital channels operating at 2400 *bits* per second. Therefore, a much greater efficiency is realized in line utilization.

Another advantage of digital transmission is that digital signals can be easily regenerated as they pass through the network. This allows error rates typically of 1 bit in 10^7 versus 1 bit in 10^5 by analog techniques. Another benefit is the reduction in complexity required for interfacing a terminal to digital transmission network.

Other techniques have been developed to transmit digital data, and examples are given under *frequency shift keying*, and *phase modulation*.

digital error An error detected as a discrepancy between a *signal* as it is transferred and received over a *line*; e.g. the displacement of a *digit* from one position to another.

digital exchange An *exchange* in which the *message* traffic is transported as *digital signals* and the connections for incoming and outgoing *traffic* are achieved by *digital switches*. ⇨ article entitled *switching equipment*.

digital filling The insertion of a defined number of *digits* into a *digital signal* to increase the *digit rate*. These inserted digits do not contribute to the information transferred, but are used to alter the timing of *signals* on a *channel* relative to certain *time slots*. Sometimes referred to as *bit stuffing*.

digital leased circuit A private *circuit* hired by a user from a *carrier* and providing a permanent connection for point-to-point *data transmission* between two *terminals* using *digital signals* in the *transmission path*. ⇨ general article on *public data networks*.

digital multiplex equipment Equipment used for combining several *digital signals* on to a single digital *circuit* using the techniques known as *time division multiplexing (TDM)*. The equipment also allows the reconstruction of the original *signals* at a distant location by demultiplexing.

digital multiplexor A *multiplexor* which operates in accordance with the principles of *time division multiplexing (TDM)*.

digital path A complete set of physical resources used to create a two-way link between two *terminals* for the transmission of *digital signals* at a specified *bit rate*. If digital paths are linked through a *digital switched circuit*, the term *digital connection* is frequently used.

digital pulses Electronic *signals* which represent information in binary coded form as a finite number of pulses; e.g. at two levels which correspond to the *binary notation* 0 and 1. Numeric information can be transmitted in this way, but also groups of digital pulses may represent alphabetic characters, numerals or special symbols. This is the primary way in which information is represented in *computers* and other electronic systems. ⇨ *binary digits*.

digital radio path A two-way *transmission path* for *digital signals* at a specified *bit rate*

54

made up of several *digital sections*, where each section comprises two radio terminal stations and their interconnecting *transmission medium*.

digital repeater An *amplifier* designed to regenerate a *digital switch* and used at regular intervals along a *transmission path* to overcome the problem of *attenuation*. ⇨ *analog repeater*.

digital section A term used in line transmission systems to refer to a segment in a *digital path* which provides a means for two-way transmission of *digital signals* between two locations at a specified *bit rate*.

digital signal An *electrical signal* made up of discrete pulses coded to represent information, and contrasted with an *analog electrical signal* which is a continuous waveform. A digital signal is of a non-continuous form and is made up of discrete pulses which take specific values; e.g. in *binary notation*, pulses representing 0 and 1.

digital sum A sum created by adding a series of *pulse amplitudes* over a given period of time to find a difference in a coded sequence.

digital switch A device for making *switched connections* between *circuits* to establish *transmission paths* for *digital data transmission*, and in which the connections are made by processing *digital signals* rather than *analog signals*.

digital switched circuit A *circuit* provided by a *carrier* for the transmission of digital *data*, which can be utilized by *users* by dialling a connection from a *calling terminal* to a *called terminal*. The circuit is only in use and charged to the user for the period involved in *call establishment*, *message transfer*, and *call clearing*.

Strictly speaking, a digital switched circuit is one using digital switching; i.e. the connections between circuits are established by operating on *digital signals* rather

than *analog signals*. In some systems, digital signals are converted to analog signals for transmission purposes. ⇨ general article on *public data networks*.

digital techniques Pertaining to the recording, transmission, and processing of information recorded as *digital signals*. ⇨ *binary digit* and *digital transmission*.

digital terminal A device connected to a *network* for the purpose of transmitting or receiving *digital signals*.

digital traffic Information transmitted over a *circuit* in which each *message* is transferred as a *digital signal* rather than an *analog signal*.

digital transmission ⇨ *digital data transmission*.

digitization The process involved in converting an item of *data* or *speech signals* into digital form for transmission, or for subsequent processing.

direct call A facility provided to allow fast set-up times in establishing a *switched connection*, in which the *network* interprets a *call request* as an instruction to establish a connection with a predetermined *address*; i.e. the selection of the address does not require a dialling operation.

direct current (DC) An electrical current in which the direction and value of the current remains constant.

direct service circuit Any *circuit* which directly links two *subscribers* or *terminals*, and is reserved exclusively for that purpose.

directional antenna An *antenna* used in point-to-point radio communication and designed to send or receive *radio waves* in a particular direction.

director exchange In the early Strowger type *telephone exchanges*, the *selectors* which set up connections were controlled directly by *signals* originating from the telephone dial. At a later stage of development, special units were introduced to translate the signals from the dial. The purpose of this development was to reduce the number of *digits* the *subscriber* had to dial. *Exchanges* adapted for this later development were called director exchanges, and the former type were known as *non-director exchanges*.

distortion An impairment of a *message signal*, caused by the characteristics of the *transmission link* and having different effects at different frequencies. For example, the *gain* or *loss* of a *signal* in a system may vary with frequency, so that the various frequencies which make up a signal are not reproduced at correct relative strengths. In an audio system, this would be observed as an impairment of tone.

Another kind of distortion is known as *non-linearity distortion*; its effect is to cause the *amplitude* of signals to be reproduced out of proportion in different *frequency bands*. It is a factor to be considered in planning *multiplexing* links which use *amplifiers* to increase the gain (i.e. signal strength) of complex signals. Unwanted harmonics are generated through the nonlinearity of the amplifiers and will cover a wide frequency band, and occur as *noise*.

To avoid the effects of non-linear distortion in *transmission links*, it may be necessary to limit *amplifier gain*, and thus position repeaters at more frequent intervals. An alternative is to compensate by the use of *equalization* techniques.

distributed data processing Many of the early *computer* systems installed in industrial organizations were based upon central computer installations, and very often *data* collected at distant locations was physically transported to the computer centre for processing. The need to link remote locations to central computers was envisaged from the beginning, but *public data networks* were not extensively avail-

able. At first, *messages* were transferred in batches at convenient times using *telex* facilities, or over *private leased circuits*, and then on-line interactive systems developed where *terminals* from remote locations could communicate with central systems over great distances.

With the advent of public data networks in the 1960s and *packet switching* in the 1970s, it became possible to install *networks* in which computer centres were linked, and large volumes of data and programs could be transferred automatically between computer centres. The idea of a single dominant central computer began to give way to a concept in which both the intelligence and the data *files* are distributed over several distant locations.

This decentralizing process provided greater responsiveness and efficiency to organizations and has become known as distributed data processing. Like many terms in the computer field, it is not always used in the same context. Sometimes the term is used simply to signify a situation where processing takes place independently at different locations, but it should be used to denote a situation in which computing power and data are shared, and where significant control and optimization problems arise.

distributed frame alignment signal A *frame alignment signal* which is distributed over a number of non-consecutive *time slots*.

distributed system Relates to any system in which *control* does not reside at one point but is distributed throughout the *network*. For example, in a *distributed data processing system*, intelligent functions are carried out at remote locations, using *terminals* with local processing power. There may be a need to communicate with a *node*, or other terminals via a node, but the terminals can function autonomously in the interests of overall efficiency, and to eliminate the dependence upon key nodes, and for the convenience of the *users*.

DTE

distribution point The point at which a number of *lines* in a *telephone system* are radiated out to different *subscribers* in a locality. An example is a telegraph pole which may be used to distribute up to 16 pairs of wires to subscribers in a certain vicinity.

distributor An item of equipment which switches *incoming calls* to appropriate *outgoing lines* in a *circuit switching* system.

In a circuit switching system, it has to be possible to connect every *incoming line* of an *exchange* to every outgoing line. For an exchange with 1000 incoming and 1000 outgoing lines, this would entail 1,000,000 *crosspoints* at which switching can take place.

More efficient internal *architectures* can be used to reduce the number of crosspoints whilst maintaining a low probability of *call blocking*. For example, a three-stage exchange could reduce the number of crosspoints in the *distributor* to 30,000. The three stages of switching entail: (a) *concentrator*, (b) *distributor*, (c) *expandor*.

The concentrator receives the incoming *local lines* and passes *calls* via a smaller number of internal links to the distributor.

The distributor switches the calls received from the concentrator or *incoming trunks* and passes them to *outgoing trunks* or to an expandor.

The expandor accepts the switched calls and makes final connection to the appropriate local lines. ⇨ *grouping*.

district switching centre A term used in the British *telephone system* to describe a *trunk exchange* at the level known as a *secondary centre*. This is somewhat confusing, since in North America the term *primary centre* is used at this level. ⇨ *exchange hierarchy*.

DLC Abbreviation of *data link control*.

DLE ⇨ *data link escape*.

document facsimile system A form of *telegraphy* in which documents, but not photo-

graphs, are transmitted over a *network*, but not necessarily maintaining the original density scale.

document facsimile telegraphy A system of *telegraphy* in which documents other than photographs are transmitted over a communications *circuit* and in which there is no guarantee of the faithful recreation of the original density scale. Compare with *photograph facsimile* and *alphabetic telegraphy*.

double current circuit A *circuit* used for *data transmission* in which voltages are applied directly to a *line* to represent *binary numbers*; e.g. positive voltage for '1' and a negative voltage for '0'. This definition is contrasted with a *single current circuit* in which a current is made to flow for 1, but is interrupted for 0. Either system can be used for transmission over short distances (a few kilometres) but, for connection to a *network* for long-distance communication, a *DCE* or *network interface unit* is needed. Such circuits are often used between a *data terminal* (*DTE*) and its DCE, when they are referred to as *interchange circuits*.

double ended control A system of *control* in which *synchronization* between *clocks* in two communicating *exchanges* (or *nodes*) is achieved by monitoring the *phase* of incoming *signals* and the phase of the clocks in each exchange.

double sideband modulation A form of *amplitude modulation* in which both the *sideband* signals produced in the modulation *envelope* are transmitted and detected. The *carrier wave* is not always transmitted in such transmission systems, and it has to be reintroduced at the *receiving terminal*. ⇨ article entitled *modulation*.

DRCS Abbreviation of *dynamically redefinable character set*.

DSU Abbreviation of *data service unit*.

DTE Abbreviation of *data terminal equipment*.

57

DTE clear request A *control* signal sent by a *data terminal* (*DTE*) to clear a *call*.

DTE waiting A *control* signal between a *data terminal* (*DTE*) and its *modem* (*DCE*), indicating that the DTE is waiting for a further control signal from the DCE.

dumb terminal A *visual display* unit used to display information and which has little or no intelligent functions to enable it to be connected to a communications *circuit*. The normal mode of use requires such a *terminal* to be connected to a local *controller*, which provides the intelligence to connect several dumb terminals to the *transmission system*.

duplex system A system of operation in which transmission between two *terminals* can take place simultaneously in both directions.

duration of a call The time interval between the completion of *call establishment* and the *call* being cleared.

dynamic equalizer A device which counteracts *delay distortion* affecting *analog signals* in transmission *circuits*, and constructed to respond appropriately at different speeds and operating conditions. ⇨ *equalization*.

dynamic port allocation Relates to any device which is capable of allocating access *ports* to incoming *channels* automatically in response to operating conditions. For example, some *high speed modems* can automatically switch channels when a channel becomes inoperative, and automatically switch channels to higher speeds of operation to make use of the full *bandwidth* on a *circuit* provided.

dynamic range The range of acoustic *signals* produced by a *message source* in terms of frequency and loudness.

dynamic test set An item of test equipment used to test the performance of *modems* and other line equipment, enabling the operator to create a wide range of operating conditions to test the efficiency of the equipment at various operating speeds and for various functions. Such test sets are usually portable and help to rapidly isolate faults in complex communication *networks*.

dynamically redefinable character set (DRCS) A technique used to enable a *terminal* to be used alternately to display different *character sets* (e.g. English, Arabic, French, etc.). In most terminals, the display character set is stored in a *read only memory* (*ROM*) where it cannot be changed readily by the *user*. Although it is possible to store more than one character set, it may not be practicable to store all the character sets possible.

A practical method is to store character sets centrally on the *databases* that the *terminal* must access. In this way, the required character set can be downloaded into terminals as required and stored in *random access memory* (*RAM*), ready to operate in a particular language.

The technique has been proposed for international systems such as public videotex systems. It is also interesting to note other possibilities; e.g. the character sets could include special symbols for engineering drawings.

E

earth network A *network* based upon installations on the ground rather than upon atmospheric reflections, and including cable or direct ground-to-ground radio links.

earth station A *transmitting* and/or *receiving station* designed to send or receive *radio signals* in the form of *electromagnetic waves* to and from the atmosphere; i.e. waves reflected from an orbiting *communications satellite*.

EBCDIC A *data communication code* much used by IBM and offering 256 unique 8-*bit* character combinations. The term is pronounced as 'ebbseedik' and is a contraction of *extended binary coded decimal interchange code*.

echo A delayed version of a *message signal* produced by reflections in the *channel*, which could impair the quality of the signal. On certain channels, particularly intercontinental telephone links, special echo-suppressing devices are used.

echo effect A condition which arises in long-distance communication over *lines*, and which appears to *users* of a *telephone* as an echo of the speaker's *voice*. It is, in fact, caused by reflections created by variations in the *circuit* (e.g. a *two-wire circuit* connected to a *four-wire circuit*). On intercontinental circuits, *echo suppressors* are often fitted to permit only one pair of wires to transmit speech at a time. If simultaneous transmission is required on such circuits, special tones have to be transmitted to disable the echo suppressors.

echo suppressor A device used on a *circuit* to suppress unwanted *signals* which are delayed versions of *message signals*, and which arise due to reflections in circuits over long distances.

editing terminal A *terminal* designed for the preparation of *text* and *graphics* by an *information provider*. For example, ⇨ *teletext editing terminal*, and *videotex terminal*.

effective call A *call* which is established so that *conversation* may take place between two people or two *terminals*.

effectively transmitted signal A definition used in the transmission of *sound programmes*, in which the *nominal overall loss* at any specified frequency must not exceed the nominal overall loss at 800 Hz by more than 4·3 *dB*. A *signal* at a frequency outside this limit is not effectively transmitted.

80 column display standard Pertaining to a *terminal* and associated transmission system which will allow a display of information containing 80 columns of *characters* on each page. For example, in a *word processor*.

electrical analog The representation of a phenomenon as an *electrical signal*; e.g. the representation of *sound waves* as *audio frequency* magnetic waves.

electrical interface The specification of the electrical requirements for communication between two devices and encompassing the *signals* which must pass between the devices to transfer *control* or *message information*. For example, the duration, frequency, *amplitude*, direction, encoding and significance of *electrical signals*.

electrical signal Any *signal* by means of which information is conveyed from one location to another, and in which the information is represented by variations either in *phase*, frequency, *amplitude* or duration of the electric current. ⇨, for example, *analog electrical signal*, and *digital signal*.

electrical telegraph This term refers to the system originally developed by Samuel Morse, in which information is represented by interrupting the electrical current in a *circuit* to transmit information as a series of pulses representing *alphanumeric* characters. The later development of this system resulted in telegraphic communication using *teletypewriters*.

electromagnetic radiation The transmission of energy resulting from charged particles undergoing acceleration, and arising as magnetic *fields* known as *electromagnetic waves* which are propagated through free space with a constant velocity of $2 \cdot 998 \times 10^8$ metres per second – the velocity of light.

electromagnetic waves Both *radio waves* and *light waves* are examples of electromagnetic waves, which can be propagated by a transmitter and *broadcast* through the atmosphere. Such waves can also be transmitted along cables or *waveguides*. Unlike *sound waves*, they can be propagated in space and need not be carried in air.

The early research into electromagnetic waves was conducted by Heinrich Hertz, who is attributed with the discovery of electric waves in the *radio frequency spectrum* and who established that such waves could pass through materials opaque to light. Today, the Hertz is a unit used to define frequency : 1 Hertz is 1 cycle per second and thus, 2 kHz is 2000 cycles per second or 2×10^3 cycles per second.

All systems of communication must generate signals which can in the end be detected by the human ear or eye. Human ears are sensitive to air waves in a frequency range from about 20 cycles per second to 17,000 cycles per second (20 Hertz to 17 kHz).

Human eyes respond to waves within a narrow band of frequency around 5×10^{14} cycles per second; i.e. frequencies of about 420 to 790 Tera Hertz, where 1 THz = 10^{12} Hz. These light waves are carried, not in air, but as electromagnetic waves which travel in space.

Radio frequencies occupy a lower frequency spectrum and the useful radio frequencies are:

low frequency (LF)	30 kHz to 300 kHz
medium frequency (MF)	300 kHz to 3 MHz
high frequency (HF)	3 MHz to 30 MHz
very high frequency (VHF)	30 MHz to 300 MHz
ultra high frequency (UHF)	300 MHz to 3 GHz

1 MHz (Mega Hertz) = 1×10^6 cycles per second

1 GHz (Giga Hertz) = 1×10^9 cycles per second

In most forms of long-distance communication, audible or visible waves are 'translated' into radio frequencies for transmission purposes, and translated back to their audible or visible forms at the *receiving station* for the *user*.

All electromagnetic waves travel through space at approximately the same speed of 300,000,000 metres per second, and the *wavelength* in metres can be calculated by dividing this number by the frequency in Hertz. (See Appendix 3.)

The wavelength determines the size of the *transmitting aerials* and *receiving aerials* required to propagate and receive *signals*. If the frequency is low approaching audibility the aerials have to be immense; if the frequencies are above the UHF *band* they are easily obstructed. Thus, the radio frequency spectrum provides a choice of characteristics which include:

LF and MF – used for radio broadcasting by the propagation of *surface waves* which are guided over the earth's surface.

HF – used for radio broadcasting by reflecting waves from the *ionosphere* – known as *sky wave* propagation. Used for long-distance communication.

VHF and UHF – both used for radio and television broadcasting as *space waves* ranging up to 300 kilometres, depending upon the geography and positioning of the

60

aerial. ⇨ *audio signals, video signals* and *radio signals.*

electron beam A stream of elementary particles of negative charge emitted from a particular source and directed at an object; e.g. emitted from the cathode of a *cathode ray tube (CRT)* and directed at a luminescent display screen.

electron gun A device used to generate a beam of electrons, a most common application being in the construction of *colour television receivers.*

electronic crosspoint A micro-electronic device used in the present generation of *electronic exchanges* for handling *voice* and *data traffic.* It consists of a simple *AND logic gate* which, when combined into a matrix with many other such devices, forms a highly efficient method for *space switching* and *time switching* of *digital signals.* Same as *electronic digital crosspoint.*

electronic digital crosspoint A device for switching *circuits* in *digital exchanges* used for *voice* or *data traffic.* The *message signals* are represented as *digital signals* and switching operations take place by synchronizing *time slots.* See general article on *switching equipment.*

electronic exchange An electronic exchange is the term applied to a *telephone exchange* which uses electronic switching components rather than electromechanical components such as *Strowger selectors* or *crossbar switches.* The evolution of electronic switches has advanced with the introduction of *pulse code modulation (PCM)* techniques, which allow *voice analog signals* to be converted to *digital signals.* Prior to this development, problems existed with exchanges which carried *analog signals;* with such signals, it is necessary to have a very high *off resistance* to prevent a signal leakage which otherwise appears as *noise* generated in the switching matrix. Electronic switches with

this characteristic are difficult to design; in addition, harmonic *distortion* is difficult to avoid.

With *binary digital signals,* these characteristics are not important, and the *crosspoint* is achieved by simple *AND gates,* one of the inputs of which is a signal input and the other is a control input. With a control signal 1 applied to the gate, the crosspoint is set on and allows the signal input to pass; with a control signal 0, the crosspoint is set off.

These *electronic crosspoints* can operate at very high speeds and allow compact switching matrixes to be created using *space switching* and *time switching* methods. The great benefits of such exchanges include a very high degree of *time sharing,* allowing *transmission paths* to be used for hundreds of *channels* at high speed, and the *network* can be expanded and controlled much more easily.

electronic mail A system providing person-to-person communication of *messages* using electronic means for entry, transmission and delivery of information in a visual form.

electronic mail box A system which allows *messages* to be placed into a storage medium, such as a *digital computer,* so that a particular *subscriber* can retrieve the message when he next *logs-on* to the system; i.e. the message is not delivered to the *user* until he logs on and requests to inspect his *file* of received messages.

element error rate The ratio of *signal elements* incorrectly received to the total number of elements sent in a given period. Also known as *bit-error rate.*

EM ⇨ *end of medium.*

emergency restart A process involving the re-establishment of a system and its *control procedures* following a major failure in which *signal* communication has failed.

emergency routes *Transmission paths*

designated for use in the event of a major breakdown of the main paths of communication.

Emile Baudot A pioneer in the field of *data transmission*, after whom the rate of *signalling* was named. Thus, we speak of transmission speeds in *bauds*, meaning pulses per second. ⇨ *modulation rate.*

en-bloc address signalling A *signalling* system in which an address is transmitted as an entity, and transmission from one stage to the next does not start until the complete *address* has been received.

encoder
1. An encoder is used in *data transmission* systems to convert *signals* within a *transmitting station* into a form required to translate physically separate signals of *synchronization pulses* and *data* into a single *serial bit stream*. At *receiving stations*, a *decoder* is situated to convert signals from the form required on the *line*, to the form needed by the *data link control* at the *receiving terminal*.
2. Any device for the generation of *characters* or *bits* to a defined *code*.

encoding law The rules established to define the *quantization levels* and their relative values, in the process known as *pulse code modulation*.

end-of-address signal A *signal* which informs the control system in a *network* (e.g. a *telephone* or *telex* system) that the preceding *characters* were *address digits* used to *route* a *call*, thus allowing the *control* to isolate and identify the particular *routing* information.

end-of-block signal A predefined *signal* or *character* which indicates the end of a *block* of *data*.

end of medium (EM) A special *character* used in a *data communication code* to signify the completion of a sequence of information recorded in *binary coded* form; e.g. the

last column from a *punched card*, or the last row in a *paper tape*.

end of message (EOM) A *transmission control* character indicating the conclusion of transmission of one or more *texts* forming a complete *message*.

end-of-message identification A *signal* or *character* which defines the completion of *data* related to a *message*.

end-of-pulsing signal A *signal* in a *telephone network* which indicates that an *address* is complete and no more *address digits* are to follow.

end of selection A similar meaning to *end-of-pulsing signal*, indicating in a *switched system* that the *address* component needed to select a *circuit* has been completed.

end of text (ETX) A *transmission control code* used to terminate a *text*.

end of transmission (EOT) An international *transmission control code* which terminates a transmission sequence and restores the *terminals* concerned to a quiescent condition.

end of transmission block (ETB) An international *transmission control code* used to denote the end of a *block* forming part of a *message*, where the overall message has been subdivided into two or more blocks for the convenience of transmission.

end-to-end layer In modern *network architectures* used for *data communications*, the system is represented as layers of logic, responsible for different functions needed in operation over the *network*. An example is given under the article entitled *ISO reference model for open systems architecture*. The end-to-end layer provides control between end *users*, including *addressing* and *flow control*.

end-to-end protocol A *protocol* which provides for the management of the *access* *path* for a *message signal* from the *transmitting station* to the *receiving station*, through any intermediate *nodes* in the *network*. May be contrasted with a system known as *node-to-network protocol*, in which access path management is from one node to the next adjacent node in the network.

engaged tone The *signal* heard by the *user* of a *telephone* when the *subscriber* being called has already lifted the *handset* to speak to another party. Also known as *busy tone*.

ENQ ⇨ *enquiry*.

enquiry (ENQ) An international *transmission control code* which is used in *polling* to enquire of a distant *terminal* its status with respect to being available to send or receive information. This code may be prefixed by *address* characters to define a specific *station*.

envelope Information recorded to encapsulate an item of *user data*, and required to ensure the effective operation of the *data network*. For example, the envelope may contain information used to control communication between *modems* or *network interface units* at each end of a *circuit*.

envelope detection *Detection* is the process by which an original *message signal* is derived from a *modulated carrier wave*. It is part of the process known as *demodulation*. In a *modulation* system in which the complete *spectrum envelope* is transmitted (i.e. no suppression of the *sidebands* or *carrier*), the method of detection is known as envelope detection. ⇨ *modulation*.

EOM ⇨ *end of message*.

EOT ⇨ *end of transmission*.

equalization A technique used to overcome *delay distortion* affecting *analog signals*. Delay distortion arises due to line characteristics which permit some frequency components of such a signal to arrive at a point ahead of other components. This phenomenon creates serious distortion of signals. An equalizer is effectively a *filter* which has characteristics that are the opposite (or inverse) of those exhibited by the *line*. Equalizers can be constructed to have different characteristics at different operating conditions, and these are able automatically to adjust to different operating speeds. They are known variously as *automatic equalizers, dynamic equalizers*, or *adaptive equalizers*.

equalizing repeaters *Repeaters* which overcome the *distortion* effects of *nonlinearity* in a *transmission channel*. ⇨ *equalization*.

equivalent bit rate The number of *binary digits* which can be transmitted over a *circuit* in a given time interval, and related to the information content of a *signal* rather than to the *control information* required to handle the signal on the particular *transmission path*. ⇨ *data signalling rate* and *data transfer rate*.

erlang A unit used to denote the utilization of a *telecommunications* system, named after a Danish engineer (Agner Erlang) who was a leader in the development of *traffic theory*. An erlang is not a precise volume of information, but a measure of the traffic experienced; e.g. an *instantaneous traffic* of 10 erlangs means that ten *calls* are in progress at a defined time. ⇨ *traffic volume*.

erroneous bit/block An erroneous *bit* is a *binary digit* signal which is not correctly received. An erroneous *block* is a unit of *data* containing one or more erroneous bits.

error bit A *bit* (*binary digit*) of information incorrectly received; i.e. the significance of the bit has been reversed – a 0 becomes a 1 or a 1 becomes a 0.

error blocks *Blocks* of *data* in which errors have been detected; e.g. by means of a *cyclic redundancy check*.

error burst A sequence of *bits* in which errors occur, and where the *error bits* are not separated by a given number of *correct bits* as required by a *circuit* of the specified standard. Also known as *burst error*.

error checking A function performed automatically in nearly all forms of *data transmission* to ensure that information transferred from one location to another is accurate. This is achieved by simple techniques, of which there are a number of variations in use.

A *binary digital signal*, for example, can be considered as a number of *coded* characters to which *parity bits* are added, so that each extended *character* thus formed will consist of a column of *bits* which must consist of an odd number of 1 bits. By checking parity along each row and column, *single bit errors* can be automatically detected and corrected.

Another form of checking requires information received to be retransmitted back to the *transmitting station* on a separate *channel* where it is compared with the original transmission.

In all forms of error checking, it is usual for the device to automatically retransmit *frames* when errors occur, and this would not be apparent to the terminal *user* unless the *line* was of very bad quality, causing transmission to be noticeably delayed. ⇨ *redundancy checking, cyclic redundancy checking, forward error correction*, and *longitudinal redundancy checking*.

error control Related to the procedures concerned with the automatic detection and correction of errors in *data transmission*. ⇨ *error checking* and *high level data link control* (*HDLC*)

error correcting code A *code* used in *data communication* which includes rules of construction so that *characters* or *blocks* of information received over a communica-

tion *line* can be checked. Departures from the rules will be detected, permitting automatic correction of some, or all, of the errors. ⇨ *redundancy checking*.

error correcting system Any system which uses an *error detecting code*, and allows for the automatic correction of errors before the *message* concerned is accepted by a *data terminal*. This may or may not involve retransmission of *error blocks*.

error correction The rules and techniques defining the way in which errors detected in *blocks* of *data* may be corrected. In most *data transmission* systems, *redundancy checking* is performed to detect errors. Correction is then performed by retransmitting *error blocks* (⇨ *automatic repeat request*) or by *forward error correction*.

error correction, forward ⇨ *forward error correction*.

error detecting and feedback system A system which uses an *error detecting code* to identify *blocks* of *data* containing errors and automatically requests retransmission of the blocks concerned. Also known as *automatic repeat request* system, or A R Q.

error detecting code A *code* used in *data communication* which conforms to rules of construction such that errors in received *data* can be identified automatically.

error detecting system Any system using an *error detecting code* but not one able to correct errors automatically. Errors may be delivered to the *data terminal* with indication of an error. Contrast with *error correcting system*.

error detection In *data transmission*, the *detection* of lost or inverted *bits* by automatic means. ⇨ *error checking*.

error peak A period during the day when the frequency of *error bits* is at its highest. For example, in systems using the *public switched telephone networks*, error peaks arise at the *busy hour* due to *impulsive* *noise* created by automatic *selectors* in *exchanges*. Systems using older mechanical selectors are particularly prone to such error peaks. ⇨ *transmission errors*.

error protocol That part of a *protocol* which deals with the *detection* and correction of errors. ⇨, for example, *half-duplex error protocol* and *full-duplex error protocol*.

error rate Transmission *circuits* are never entirely free of *noise* and other impairments, and it is expected that errors resulting in the corruption of *bits* will occur from time to time. Circuits are constructed to conform to international standards and the *error rate* is one of the factors used to designate the quality of a circuit. The error rate is given as an average rate for the occurrence of errors for a given volume of bits transmitted; e.g. 1 error in 10^7 bits transmitted. Equipment used to transmit and receive digital information contains *error checking* facilities to seek automatic retransmission of *frames* containing errors.

ESC ⇨ *escape*.

escape (ESC) ⇨ *escape code*.

escape code A *character* in a *data communication code* which changes the meaning of a group of ensuing characters. The use of an *ESC* character extends the range of meanings possible from a finite set of character codes.

ETB ⇨ *end of transmission block*.

Ethernet This system was one of the first products intended as a basis for *local area networks*. Ethernet was developed jointly by Digital Equipment Corporation, Intel Corporation and the Xerox Corporation and based upon prototype work at Xerox. It is included here as an example of the objectives sought in such networks, but other products have followed different principles.

Ethernet is intended to provide for communication between up to 1024 *stations*, at a *data rate* of 10 million *bits* per second, over distances around 2·5 kilometres.

The *transmission path* is provided by a *co-axial cable loop* to which all *user* terminals are *interfaced*.

Its specification includes the definition of the two lowest levels of an overall *network architecture* up to what is known as the *link level* facility. Thus, higher level facilities are to be provided by structures built into the user devices (*computers* and *terminals*). Local area networks of this type exhibit very low *error rates* and their main objective is to carry *bursty traffic* at *high peak data rates* between a group of users wishing to share facilities in an economical way.

The users all share the same transmission path, and communication is achieved by loading information with appropriate *address* codes for the user destination on to the *cable loop*, with special features designed to prevent *collision* of data and to ensure that over a given time all users have equal access to the *network*.

ETX ⇨ *end of text*.

Euronet This is a European-wide *packet switching* service, which has been developed by the European Economic Community (*EEC*) with the co-operation of the *PTTs* of member countries. It is based upon an implementation of *X25* and includes 5 *packet switching exchanges* (*PSE*) located in Frankfurt, London, Paris, Rome and Zurich. There are also 5 *remote access points* (*RAP*), located in Amsterdam, Copenhagen, Brussels, Luxembourg and Dublin, and a *network management centre* (*NMC*) in London.

The system is designed to meet the growing need for public *data communication* services in Europe and for European-wide *information retrieval* services.

even parity A condition which is said to exist when all the *bits* in a particular row or column of a *block* of digital *data* add up to an even number. *Parity checks* are made as part of standard *error checking* procedures and, where even parity checking is used, the occurrence of *odd parity* signifies an error in a block of received data.

exchange An exchange is the element in a *telecommunications* system which controls the *traffic* to and from destinations. It allows any *user* to communicate with any other within a large distributed population of users.

The most modern exchanges are capable of carrying *voice*, *data* and *telex traffic*, but the majority of exchanges in use today are still designed specifically for one kind of traffic; e.g. *voice traffic* over a *public switched telephone network*, or data over an international *telex* network.

Within any particular system, there may be a number of exchanges performing specialized functions; e.g. in a telephone network *local exchange*, *trunk exchange*, *primary exchange*. ⇨ *exchange hierarchy*.

In North America, the term *office* is used instead of exchange. ⇨ general article entitled *switching equipment*.

exchange hierarchy A *network* arrangement including the interlinking of *trunk exchanges* by *transmission paths* to provide a *subscriber switching* service. In any communication system involving a large number of *users*, it would be impractical to have a separate pair of wires to connect each *terminal* to every other in the system. If one was to attempt to construct a system in this way, it would be necessary to provide $(N(N-1) \div 2)$ separate line pairs, where $N =$ the number of terminals.

In modern communication systems, it is usual to establish a number of central switching points (*exchanges* or *nodes*), which have high capacity *transmission links* (known as *trunks*) between them. These *trunk circuits* are normally *multiplexed* so that a particular circuit can handle several hundred *messages* at any one time; i.e. these links are shared by many concurrent *calls*.

The terminals in the system are all wired into a particular exchange by *local lines* and each terminal thus has a specific *transmission path* to its own *local exchange*. In the *telephone system*, a *subscriber* is usually directly connected to his local exchange when he lifts the *handset* from his terminal. When the user dials a number, he generates *codes* which are interpreted by the local exchange. If the subscriber being called is on the same local exchange, the dialling signal is switched automatically to call the appropriate terminal. If the subscriber called is not on the local exchange, the local lines from the *calling terminal* are automatically switched to a trunk exchange.

Trunk exchanges carry out the necessary switching to *route* messages which must pass through the *trunk network*. It is not usual to have a direct connection between every trunk exchange. To fully interlink N trunk exchanges in a network to every other would require $(N(N-1) \div 2)$ trunks. In practice, an exchange hierarchy usually exists, which provides a more economical use of trunks.

In such a hierarchy, every local exchange has a parent exchange to which it is connected by a *trunk link*. These are known as the *primary trunk exchanges*. Primary trunk exchanges may be interlinked with other primary trunk exchanges, but only to those with which it is expected to handle a high *traffic volume*. The connection to other primary trunks would otherwise be through a *secondary trunk exchange*, and maybe beyond this, through *tertiary* or *quaternary trunk exchanges*.

The exchange hierarchy can be represented as an inverted tree structure as shown in Appendix 12.

The exchanges at the top of the hierarchy are fully connected, each to every other at that level.

There are many possible routes for a particular call, and the combination of trunks and exchanges to make a call is known as its *routing*.

The routing which uses the least number of *trunk routes* is known as the *basic routing* for a call; other routings are known as *alternative routings*.

expandor An item of equipment which accepts *calls* on the internal links of a *circuit switching exchange* and passes them to the appropriate lines. ⇨ *concentrator* and *distributor*.

extended control set A set of *control* codes (e.g. *cursor control*) which provide for *transmission control* and display functions over and above the *codes* contained in the basic set of a *data communication code*.

extension circuit Any *circuit* which is used to provide a permanent connection to connect a *terminal*, or *node*, to some other facility or location; e.g. to give access to a particular *network*.

extension codes A range of *characters* which can be available as an extension of a basic *character set* in a *data communication code*. Here the basic set represents the minimum requirement for compliance with a particular code standard.

F

facility request A part of a *signal* to be sent over a *public data network* and used to select the *network facilities* required as part of a unit known as the *selection signals*.

facsimile A branch of *telegraphy* concerned with the reproduction of images over a *telecommunications* system. The reproduction is in a permanent form (e.g. recorded on paper), and may include photographs, monochrome images, colour or intermediate shades. The term *document facsimile* is sometimes used to imply the reproduction of documents without the ability to recreate photographic images, and the term *photographic facsimile* is applied in the latter case.

facsimile apparatus A device used to transmit documents by *facsimile* methods and which may, or may not, use special techniques to compress the *bandwidth* of the *signal* to achieve a particular transmission time. Such devices are rated according to the transmission time into groups. The transmission time is based upon a standard *ISO* A4-size document using the public telephone network as a *transmission medium*; e.g.

Group 1 apparatus 3 to 6 minutes
Group 2 apparatus 3 minutes
Group 3 apparatus 1 minute

Such devices use *double sideband modulation* and, in the case of Group 3, include means for reducing redundant information in the *signal*, prior to *modulation*.

facsimile baseband The name given to *electrical signals* generated directly as an output from the scanning process in a *facsimile* system. These signals constitute the primary *message signal* which is usually *modulated* on to a *carrier wave* for transmission.

facsimile telegraphy A system of *telegraphy* which permits the reproduction of images in permanent form over a communications *network*. Each image is captured by a scanning process which generates *electrical signals* to represent the image. These signals are transmitted to allow reproduction (e.g. as a photograph or printed page) at a distant *station*.

facsimile transceiver A device used to transmit or receive *signals* using the techniques of *facsimile transmission*.

facsimile transmission More strictly known as *facsimile telegraphy*. *A method* for transmitting and receiving *graphic* images to reproduce an image at a distance. The process entails scanning a picture or image to create *electrical signals* which can be transmitted over a *network*. These signals are often known as the *facsimile baseband* and, in many systems, *modulation* techniques are used to transmit the signals on to a *carrier wave* to a distant *station*. At the distant station, a *facsimile transceiver* demodulates the incoming *signal*, and the derived signal is used to modulate the intensity of a light source in order to reproduce the black and white shades of the original image. The technique has also been adapted to the reproduction of photographic images containing continuous tonal densities which are faithfully reproduced. This is known as *photograph facsimile telegraphy*.

The transmission of images other than photographs, entailing a lesser degree of accuracy in the reproduction of the density scale, is known as *document facsimile telegraphy*.

fast circuit switching A method of providing *data transmission* facilities in which a *user* dials a connection to another, using a *data circuit* provided by a *common carrier* for the duration of the *call*. The availability of such circuits (providing *call set-up times* in fractions of a second and *billing* increments in fractions of a second) provides attractive economic solutions to intermittent communication requirements where a *leased circuit* would be too costly. ⇨ *public data networks*.

fault Any condition in equipment or *software* which causes a system to operate below the level of service specified for *users*.

fault detection ⇨ *fault diagnosis*.

fault diagnosis The practice and procedures involved in determining the particular conditions causing errors or interruptions to a communications service. Usually the responsibility of an engineer or *network manager*, but often aided by complex test equipment which is able automatically to detect component failures.

fault rate An expression used in assessing the reliability of a system, or system component, and referring to the average number of faults occurring in a given period of time. ⇨ *mean down time*.

fault report point The location or organizational entity to which faults in a system are to be notified. Sometimes highly automated procedures using special *circuits* are provided.

FCS Abbreviation of *frame check sequence*.

FDM Abbreviation of *frequency division multiplexing*.

feedback A term used in process control in which some part of the output produced by a system is monitored and used as an infeed to the system to regulate its operation or performance. e.g. in certain electronic amplifier systems, the output signal may be added or subtracted from the input signal to maintain the output stability within a desired output range.

FF ⇨ *form feed*.

fiber optics A branch of communications technology in which information is transmitted as light pulses along specially constructed fibers. The fibers are made of a central core bounded by a sheath of material with a much lower refractive index. Light signals applied at one end of the fiber, are conducted along the core because the light is reflected from the outer sheath. The fiber is said to exhibit total internal reflection.

The individual fibers can be of very small diameter, and they may be bundled together to create flexible cables in which each fiber may provide a broadband channel. ⇨ *optical fibers*.

field A unit of *data* in a *record* or *message* and designated for a particular purpose. For example, an *address field* containing *routing* information, or a data field used in a particular application such as job number, staff number, quantity in stock, name and address field.

field flyback An interval during the scanning operation of a television picture, in which the *electron beam* is blocked off and the *raster* is repositioned to begin a new scanning operation. ⇨ *video signal*.

field frequency The frequency with which a complete sequence of scanning operations takes place in the creation of a television picture. Measured in cycles per second or *Hertz* (Hz): typical frequencies are USA 60 Hz and *CCIR*-625 lines system 50 Hz. ⇨ *interlaced scanning* and *video signal*.

field scan The vertical movement of the *raster* during the scanning operation to create a television picture. ⇨ *video signal*.

15-supergroup An assembly of equipment carrying fifteen separate *supergroups* in a *frequency division multiplexing* system and providing a means of transmission for up to 900 separate *speech channels* along a single *co-axial tube*. Also known as a *hypergroup*.

50 per cent duty cycle In *time division multiplexing* systems, *digital signals* are transmitted as waveforms which, due to line conditions, become rather more rounded than the square pulses originated by the digital source signal to represent 0's and 1's. Sometimes it is arranged that pulses delivered to the *line* occupy only half the duration of the *time slot* allocated to the *digital pulse*. This creates a sharply defined waveform which assists in the identification of pulses.

figures shift A special *character* in a *telegraph code* or *data communication code* which has the function of designating all subsequent codes in a transmission sequence as being numbers (0 to 9) or characters from a secondary group of codes. The figures shift remains in force until a *letters shift* occurs.

file An accumulation of *records* in a *digital computer* or on any device capable of storing digital information. The file will be structured to permit individual records or groups of related records to be selected and processed. Records may be selected or sequenced according to the content of *fields* within the records.

file separator (FS) A special *code* under the general classification known as *information separators*, and used in a *data communication code* to denote the boundary of a specific *file* of *binary coded information*.

filter A device used to control the frequencies which can pass along a *circuit*, enabling selected frequencies to pass without significant *attenuation* but blocking frequencies that are not desired. ⇨ *low pass filter*, *band pass filter* and *high pass filter*.

first exchange Refers to the *exchange* closest to a *calling party* or to the first exchange in any connection which utilizes a specific *signalling* system; e.g. the first *No. 6 exchange*. ⇨ *last exchange* and *intermediate exchange*.

first-party clearing A method of *call clearing* in which the *call* is cleared when either party sets down his *handset* on the *switch hook*. Contrast with *calling party clear*.

5-bit code A *data communication code* in which each *character* is represented by a combination of five signal elements, and thus allows 2^5 (or 32) characters, each formed of unique code combinations, to be used.

fixed path protocol (FPP) A method used to transmit information over a *packet switching network* in which a *virtual circuit* is established such that all *packets* relating to a *call* use that identical path. The path is released when either party hangs up. Contrast with *path independent protocol*.

fixed reference phase modulation A method of *phase modulation* in which the *binary digits* 0 and 1 are indicated by a *unit signal element* beginning with a *phase inversion* in a specific direction. With this method, a *reference wave* is required in the *demodulator* to provide *detection* of the *phase* of the *data signal*.

fixed virtual circuit Same as *permanent virtual circuit*.

flicker effect An effect perceived by a person when he looks at a television picture (or some form of visual display), in which the interval between successive pictures being presented is too long, relative to the rapid decay of *luminance* generated by the *electron beam* on the tube surface. The frequency below which this effect becomes apparent varies with individual people, but a *field frequency* of 60 per second is used in North America and 50 per second is common in Europe. ⇨ *interlaced scanning*.

floppy disc A medium for storing digital information and consisting of a disc similar to an 8″ gramophone record, upon which information is recorded as magnetized *fields*. A floppy disc is so called because the material from which it is made is not rigid. Such discs are readily handled by operators and are in common use for small office *computers* and *word processors*.

floppy disc drive A mechanism for handling *floppy discs* during the operation of a system, consisting of a drive mechanism to revolve the disc, and magnetic read/write heads which record or retrieve information.

flow control The process which takes place in a *data network* to regulate the transfer of *data* between any two points. Since different *data terminals* (*DTE*) and *nodes* in a network may operate at different speeds, *control* has to be exercised to store *frames* or *packets* at various points to prevent *buffers* at other points becoming overloaded.

flow control information In any *packet switching* or *message switching* system, the rate of packet flow has to be regulated so that *buffers* at *receiving stations* or intermediate *nodes* do not overflow. To avoid overloading parts of the system, queues may be formed at intermediate buffers, and alternative *access paths* may be chosen to avoid bottlenecks. To achieve *control*, flow control information has to be fed back along the system to advise the *transmitting stations* of the current capacity and status of *receiving stations*.

fluorescence Light radiation produced from certain materials; e.g. the light energy produced when an *electron beam* strikes phosphor materials coated on the surface of a TV tube. ⇨ *phosphorescence*.

flyback An event during the scanning operation which takes place to create a television picture, and relating to the movement of the *scanning beam* back to a point

at which the next scanning movement can commence. There are two kinds of flyback, *line flyback* and *field flyback*. ⇨ *video signal*.

flyback periods The intervals of time during which the *scanning beam* is blocked out to enable it to be repositioned to begin the next *line* or *field scanning* operation in a television system. ⇨ *video signal*.

FM Abbreviation of *frequency modulation*.

FM broadcast A radio broadcast using the techniques of *frequency modulation*.

form feed (FF) A function represented as a special *format effector* in a *data communication code*, and serving as an instruction to move a *print mechanism* or the *cursor* of a *visual display unit* ready to print or display a new form.

format effectors Special symbols forming part of a *data communication code* and used to control ancillary devices such as printers connected to a communications line. Examples include:

BS back space
CR carriage return
FF form feed
HT horizontal tabulation
LF line feed
VT vertical tabulation

40-column display standard Pertaining to a *terminal* and the associated transmission system which will allow up to 40 columns of *characters* on a single *page* of information. Examples include *videotex* and *teletext* systems.

forward channel A *channel* in a *data network* in which the direction of transmission is from a *transmitting terminal* to a *receiving terminal*. This is to be contrasted with a *backward channel* in which supervisory information and *control* signals will flow in the opposite direction to the *user* information on the forward channel. It

71

should be noted that *data communication* systems can transfer user information in both directions, and the terms 'forward' and 'backward' channel are related to the *data source* at any instant.

forward direction A term used in *telephony* and in *data communication* to denote the direction of information flow, as distinct from the *backward direction* in which supervisory and *error control* information may flow. It is important to note that information may flow in either direction, and so this term is used relative to the information flow at any instant.

forward error correction A method of *error detection* and *correction*, based upon *redundancy checking* techniques, in which a high volume of redundant *bits* is computed and added to transmitted *data*. The redundancy is such that the *receiving terminal* is able to automatically correct and not simply to detect errors.

Fourier components Fourier was an early researcher into complex waveforms, and he determined that each such waveform is composed of fundamental frequencies and harmonics which are exact multiples of their fundamental frequency.

In dealing with the design of physical devices to handle such waveforms, engineers are able to perform an analysis to identify these so-called Fourier components and deal with them separately on the basis of simple sine wave theory. Fourier analysis is a technique widely used in communication involving *electrical signals*.

four-wire channel A *transmission path* in which two pairs of wires are used, one pair for each direction of transmission. Strictly speaking, a *channel* provides a one-way path and the term *four-wire circuit* is more correct.

four-wire circuit A *transmission path* between two *terminals* in which there are two separate *channels*. For example, in *data*

communication, one channel may be used to transmit information and the other may be a return channel used for supervisory or *error control* data. Four-wire circuits are also used to provide for *multiplexing*, or *amplification* of *signals* on *speech channels*.

four-wire link This form of *link* refers to a *transmission system* which uses separate pairs of wires for each direction of transmission. A four-wire link is usually created to allow *amplification* in an *analog* circuit or to permit *multiplexing* of several *channels* on to a single physical channel.

FPP ⇨ *fixed path protocol.*

frame
1. Generally used in *data transmission* to denote a unit of information used in transmission, and upon which various automatic *error checking* operations are performed to ensure correct and error-free transmission.
2. A frame in a *videotex* or *teletext* system is equivalent to one screen display of information, and there may be 26 such frames to a designated *page*.
3. In television broadcasting, a frame is a unit of transmission equivalent to a screen; e.g. in the United Kingdom, 625 lines make up one picture or frame.
4. In a *time division multiplexed* system, a set of *time slots* in which the *digit position* of each time slot is defined in relation to a *frame alignment signal*.

frame alignment A process in which the timing of a *receiving terminal* is correctly adjusted to the alignment of *frames* in a received *signal*.

frame alignment signal A part of a *message signal* which serves to identify the boundary of a *frame*, and assists to create *synchronization* between a *receiving terminal* and an incoming *signal*.

frame check sequence A sequence of *bits* attached to a *frame* to provide for the *detection* and correction of errors in a *data link*. ⇨ *cyclic redundancy check.*

frame format A definition of the structure of an indivisible unit of information transmitted in a *data communication* system, including *destination address*, *source address*, frame type, the *data*, and *frame check sequence*. The structure of the *frame* is an important part of the specification for a *data link*.

frame grabber An element in a *receiving station* which inspects *address information* in *frames* passing along a *channel*, to detect whether the frame should be received by the particular *station*. This technique is sometimes used in *cable loops* used for *local area networks* where all *transmitting stations* and receiving stations may be connected to the same channel, and where all frames transmitted include *destination addresses*. It is also used in *teletext* systems to grab specific frames transmitted as part of a general *broadcast*.

frame level functions Processes performed in a *packet switching system* to check the transmission of *data*, to ensure that *line* errors have not occurred and to retransmit *frames* automatically when errors are detected. It is the main purpose of the *high level data link control (HDLC)* to be responsible for these functions.

frame level protocols In *data communication*, using *high level control functions*, *control* is exercised by procedures and rules which enable small units of information known as *frames* to be transmitted and received securely with automatic *error checking*. The control procedures exist at various levels and those concerned with the identification sequencing, *routing*, and checking of frames, are referred to as frame level protocols. ⇨ *packet switching network* and *X25*.

frame store A *memory unit* capable of storing a display of information for a *page* (or *frame*) of *text* and *graphics*. For example, the memory unit in a *videotex terminal*. Also known as *page store*.

frame (videotex system) In *videotex systems* as exemplified by the British system known as *Prestel*, the *database* consists of individually numbered *pages*, each page being the equivalent of one screen display of 960 *characters*. Each page can include up to 26 continuation pages (identified by letters a to z); each such continuation page is termed a *frame*.

framing bits Extra *binary digits* attached specifically to *time division multiplexed* signals to provide *control information* for the correct transmission and separation of individual *frames*. The framing bits identify the individual *terminals* concerned with particular *messages*, and also identify the start of message *characters*. ⇨ *time division multiplexing*.

frequency band A range of frequencies that can be propagated through a particular *circuit*, or a range assigned for any designated purpose. For example, *radio frequency signals* exhibit different characteristics according to the point in the *frequency spectrum* at which they occur. *Signals* in the *high frequency* band (*HF*) from 3 MHz to 30 MHz are able to be deflected off the *ionosphere* and can be bounced around the world. They are suitable for all kinds of long-distance radio communication.

Electromagnetic waves higher than 30 MHz are not reflected and can only travel between points in direct line of contact. They are suitable for short-distance radio communication; e.g. television, *radar*, radio broadcast communication.

Within any particular frequency band there are further subdivisions or *bands* allocated for specific purposes. For example, in the United Kingdom, the various *television channels* in the *ultra high frequency* (*UHF*) spectrum occupy frequencies from 470 MHz to 940 MHz. ⇨ *frequency spectrum* and *bandwidth*.

frequency division multiplexing (FDM) In a *multiplexing* system, a single *channel* of suitable *bandwidth* is used to carry a number of *transmission channels* of narrower bandwidth. In frequency division multiplexing, a technique which is used for analog *speech channels*, this is achieved by modulating each *message signal* on to a separate *carrier wave*. Thus, the resulting signals are separated from one another by a frequency displacement which ensures that there is not mutual interference.

In a typical FDM system for transmission of telephone signals, each channel is created by means of *single sideband modulation* of separate carrier waves, and the resulting signals are combined. In the modulation process, the message signals are limited by means of *filters* to a bandwidth of 3 kHz. The carriers are separated in steps of 4 kHz to provide an assembly of 12 channels known as a *group* in the *band* from 60 to 108 kHz. The group, thus, has a bandwidth of 48 kHz.

A second stage of multiplexing allows five groups to be used to modulate carriers in steps of 48 kHz between 420 kHz and 612 kHz. The *lower sidebands* are selected to provide an assembly of 60 channels known as a *supergroup* in the band 312 to 552 kHz; i.e. a bandwidth of 240 kHz.

Further stages of multiplexing to combine supergroups may take place. For example, a combination of 900 channels created in this way is sometimes referred to as a *hypergroup*. When the groups have been assembled into a single band, they are transmitted over the *transmission path* (e.g. a *co-axial tube*), and then *demultiplexed* by a reverse process.

The number of channels which can be carried over an FDM channel in this way depends upon the *transmission medium*. The maximum *frequency band* of a *multiplexed channel*, using 2·6/2·9 mm co-axial tube with *repeaters* staged at 4-kilometre intervals, is in the region of 12 MHz. This would sustain approximately 2700 *telephone channels* using the techniques mentioned above.

This form of multiplexing is known as frequency division multiplexing because each signal occupies a different frequency band when transmitted. For the transmission of *digital signals* a different technique known as *time division multiplexing* (*TDM*) is used.

frequency error An error arising to affect the stability of the frequency of a *signal* or *carrier wave* in a *multiplexing* system. Usually this sort of error must be kept within a *band* of ±2 Hz. Also known as *frequency offset*.

frequency modulation (FM) A method for carrying a *message signal* imprinted upon another *carrier signal* which is at a selected frequency to suit the particular mode of transmission. The technique is used in *radio transmission*, also in transmission of telephone signals over a *transmission path* used to carry many simultaneous messages. ⇨ *modulation* and *frequency division multiplexing*.

frequency offset An error condition affecting the stability of the frequency of a *signal*. For example, the *carrier signals* in a *multiplexing* system. Such frequency variations have usually to be contained to ±2 Hz. Also known as *frequency error*.

frequency response The ability of a particular *circuit* or device to carry *signals* of different frequencies is known as its frequency response.

frequency shift keying (FSK) A method of *frequency modulation* used for the transmission of binary information, in which the frequency of a *carrier signal* is changed to represent the *binary digits* 0 and 1. The occurrence of several contiguous cycles at a particular frequency equals 0, and the occurrence of a similar set of cycles at another frequency equals 1. Thus, the carrier will be seen to vary between two frequency levels, without a change in the *phase* of the signal. ⇨ *unit signal element* and *modulation*. Compare also with *phase modulation*.

frequency shifting A technique used so

that several thousand *calls* can be carried simultaneously on the same *transmission channel*. This process is achieved by *multiplexing*; i.e. each *message signal* is modulated on to a different *carrier signal* for transmission, and the frequency of each carrier signal is chosen such that each multiplexed signal occupies a different *frequency band* from all the others. ⇨ *modulation* and *frequency division multiplexing*.

frequency spectrum *Electromagnetic waves* and *light waves* can be grouped into *bands* which provide convenient classifications for the purpose of considering their characteristics for propagation and use as communication vehicles. Some examples are listed below:

Voice frequency – the band of frequencies used to carry a human voice *analog signal* in a telephone *network*. From 300 to 3,300 cycles per second (300 to 3,300 *Hertz*).

Audio frequency – the band of frequencies used to carry an analog signal equivalent to the range of sounds perceived by human ear. From 20 cycles per second to 20,000 cycles per second (20 kilocycles, written as 20 kHz).

The *radio frequency* spectrum includes a great range of frequencies, among which the more useful are in the range from 20,000 to 20,000,000,000 cycles per second. All forms of electromagnetic and light waves travel through space at the same speed of 300,000,000 metres per second. In radio broadcasting and communications, it is useful to consider the following frequency bands:

low frequency (*LF*). A range of signals from 30 kHz to 300 kHz. They are suitable for long-distance radio communication and are often used for military or trans-oceanic services. At the lower frequencies, a large *antenna* is needed to propagate and receive signals.

high frequency (*HF*). A range of signals from 3 million cycles per second to 30 million cycles per second, 3 to 30 MegaHertz (MHz). Used for long-distance communication, but the quality is dependent

upon ionization in the upper atmosphere.

very high frequency (*VHF*). A range of signals from 30 MHz to 300 MHz. Used for short-distance radio communication.

ultra high frequency (*UHF*). A range of signals from 300 MHz to 3,000 MHz. Often used for television broadcasts, and covering *frequency bands* from 470 MHz to 940 MHz.

super high frequency (*SHF*). A range of signals from 3,000 MHz to 30,000 MHz, otherwise expressed as from 3 GigaHertz to 30 GigaHertz (GHz). ⇨ *frequency band* and *bandwidth*.

front-end processor A computer subsystem used mainly to *interface* a main *computer* or *host processor* to a communication *network*. It takes responsibility for the *communication control* activity rather than the *application programs* which are run in the host system. ⇨ *communications controller*.

FS ⇨ *file separator*.

FSK ⇨ *frequency shift keying*.

full availability transposition It is an objective in designing *exchanges* to minimize the number of *crosspoints*, whilst at the same time reducing the probability of *call blocking*. A method to achieve this optimization consists of arranging the crosspoints in stages in which a number of matrixes known as switching groups occur. Full availability transposition is said to exist when the pattern of interconnection is such that every *group* in one stage has a connection to every group in the previous stage.

full duplex A *transmission channel* in which simultaneous two-way transmission is available. ⇨ *half duplex* and *simplex*.

full duplex error control A system of *error correction*, used in *data transmission* over *links* which have a long transmission delay time, such that a *transmitting terminal* transmits a whole series of *blocks* without waiting for the *receiving station* to acknowledge correct acceptance of each block separately. If an error is detected, individual *error blocks* are retransmitted or all blocks commencing with an error block are retransmitted. These techniques require blocks to be numbered, and the latter method is also referred to as *go back to N* technique.

full echo suppressor An *echo suppressor* in which the *speech signals* on each path are used to control the suppression loss in the other path of a *four-wire circuit* used for long-distance communication. Contrast with *half echo suppressor*.

fully provided route A *transmission path* designed to handle all the *offered traffic* without relying on any *alternative route* in times of *peak load*. ⇨ *high-usage route*.

functional compatibility Most modern communication systems are of modular design; i.e. it is possible to extend the capacity of the system by adding additional equipment components. A new item of equipment must be capable of supporting all the logical functions of the existing system, and must also exhibit *interface compatibility* (e.g. electrical and physical compatibility).

functional specification A document describing the attributes of an item of equipment and/or *software* system. The emphasis is on what the equipment does, rather than how it does it, and should explain how the inputs and outputs are related.

functional test A test to determine whether a device or a *circuit* will operate correctly under working conditions to fulfil the specification for the quality of service defined for a system.

G

gain The term used to describe the increase in the power of a *message signal* during the process of *amplification*. In contrast, we may consider the *loss* of power in a message signal passing through a cable.

Amplifiers are introduced into communication *circuits* to overcome cable losses, and to deliver message signals to the message destination at an appropriate *minimum signal level*.

gaussion noise A form of interference which manifests itself as a hiss on telephone or radio *channels*, and is caused by the natural movement of electrons in *circuits*, which varies with temperature. Also known as *white noise* and *random noise*.

generating polynomial A mathematical expression used in the creation and checking of *cyclic codes* used for *error checking* in *data networks*. The method relies upon any element of *data* being regarded as a pure *binary number*. The generating polynomial is used as a means of calculating a remainder which is attached to each *binary string* prior to transmission. The same formula is used at the *receiving station* to derive the remainder once more, and a check against the transmitted remainder is performed.

go back to N A *transmission control* signal which acts as an instruction to repeat a transmission from a specific *block* of *data*.

go path A *transmission path* from a *terminal* or *node* used for outgoing *signals* and *messages*. Compare with *return path*, which relates to a *channel* used to transmit supervisory or *error control* information.

grade of service A measure of user service expressed as a probability. For example, on lifting a telephone receiver,

the probability that the *dial tone* will not be received, or the probability that an *engaged tone* may be received due to system congestion. ⇨ *probability of call blocking* and *probability of excessive delay*.

graphic character In some forms of graphic presentation, illustrations to be displayed upon screens are realized by combining individual *characters* to form an overall picture. These characters are themselves composed of a number of individual dots (*picels*) which, when resolved together on the screen, form the shape of the character. An example is provided in the *alphamosaic graphics* used in certain *videotex* systems.

graphic display standards Pertaining to the way in which *graphic information* is stored and transmitted in a system to be displayed upon a screen. The possible fineness of the illustrations is governed by the techniques adopted within the display terminal. Some standards do not allow for fine illustrations and are known as *low resolution* display standards; others do allow for fine illustrations and are known as *high resolution* display standards. ⇨, for example, *videotex standards*.

graphic information Digitally encoded information which represents a description of a drawing. In many *graphical processing* applications, the information actually transmitted is highly condensed to provide for rapid transmission of a detailed picture. For example, straight lines and corners are represented by reference co-ordinates and the interpretation and plotting of the desired pictures is achieved by programs in the *receiving terminal*.

graphical processing Some forms of communication allow drawings to be conveyed over *circuits*. This function is important in training, management and operational applications. A variety of devices are used, but essentially it must be possible to scan a drawing displayed upon a pad or display screen and to create *digital signals* representing the features of the illustration. These signals are transmitted and retained in a store which is a map of the display screen at the *receiving terminal*. The store at the receiver can be used to refresh the screen so that stored pictures can be displayed for as long as required.

graphics Any *data*, presented to end *users* as an illustration rather than as pure *text*, may be referred to as graphics. Certain *computer* applications are of value only if data in the form of illustrations, charts or graphs can be presented, along with text, to end users.

ground waves Radio broadcast signals used in short-range *high frequency* communication, consisting of *surface waves* plus waves reflected from the ground.

group In relation to *frequency division multiplexing*, an assembly of equipment providing a means of transmission for 12 separate *voice channels* along a single *transmission path* with 48 kHz *bandwidth*.

group delay/frequency response An effect experienced when transmitting *data signals* over *telephone channels*. This term is used to describe the difference in the delay of different frequencies due to circuit conditions which affect the propagation time of different frequencies. The delay factor is dependent upon the signal frequency and the length of the *circuit*. This phenomenon is not generally noticeable to a *user* in *speech communication* but may be significant in *data communication*. The problem can be overcome by the use of *delay equalizers* which effectively slow down the advanced signals.

group link The complete means of transmission for a group of 12 *speech channels*

using a *bandwidth* of 48 kHz over a *network*. ⇨ *grouping*.

group 1/2/ or 3 facsimile apparatus ⇨ *facsimile apparatus*.

group switching centre A term used mainly in the United Kingdom to describe a *telephone exchange* which receives *trunk lines* from several *local exchanges* and *interfaces* to other trunk lines which connect to *secondary trunk exchanges*, forming the *trunk network*. Known as a *primary centre* (CCITT) or *toll center* (USA).

group separator (GS) A special *code* under the general classification known as *information separators*, and used in a *data communication code* to denote the boundary of groups of *records* in binary coded form.

grouping This term is used to define the way in which a number of separate *message signals* can be combined for transmission over a single *co-axial tube* using techniques of *frequency division multiplexing*. Each message signal is used to modulate a different *carrier wave* using *single sideband modulation* and the derived signals are combined into groups. Usually groups of 12 *channels* are chosen, each occupying 4 kHz within the *band* from 64 kHz to 108 kHz.

Further stages of multiplexing may be used to combine groups in 48 kHz steps within the band from 420 kHz to 612 kHz, resulting in a 60 channel group known as a *supergroup*. Further extensions of the grouping principle allow 16 supergroups to be combined to form a *hypergroup* of 960 channels.

GS ⇨ *group separator*.

guard band An unused portion of a *frequency band*, used to separate different *channels* in the *bandwidth* to prevent mutual interference between adjacent channels.

gun ⇨ *electron gun.*

G1 code set A set of *graphic characters* forming part of a standard for display of *alphamosaic graphics* in a *videotex* system.

G2 code set A set of *characters* and their means of representation in a *videotex* system, and covering a range of special symbols and diacritical marks used in different languages.

H

half duplex
1. Sometimes used to indicate a *transmission channel* in which two-way transmission is available but only one way at a time. But see *simplex* 2.
2. The *CCITT* definition describes a half duplex *circuit* as allowing two-way transmission, but the *terminals* connected can only receive or transmit at any one time. ⇨ *full duplex*.

half duplex error protocol A system in *data transmission* in which the *transmitting terminal* waits for a response from the *receiving terminal* after a *block* has been transmitted. This is done so that the transmitting terminal may know whether to retransmit the block because of an error, or to send the next block. Compare with *full duplex error protocol*.

half echo suppressor An *echo suppressor* in which the *speech signals* on one path only are used to control the suppression loss in the other. Contrast with *full echo suppressor*.

Hamming distance The number of *digit positions* which differ between two *words* of the same length and radix. For example, in 3789124 and 2719124 the Hamming distance is 2.

handset That part of a telephone *terminal* which the *user* lifts to speak and hear through when making a *call*.

handshake A procedure followed to interconnect two *data terminals* to initiate a *call* prior to an exchange of information known as a *conversation*. The handshaking procedure effectively establishes the *transmission* path for the *call*, and is also known as *call establishment*.

hang-up signal A *signal* sent to an *exchange* to indicate that a *called party* has cleared by replacing his *handset* to terminate a *call*. The exchange must then clear the call and stop the charging process.

hard copy Describes any form of output from a system in which *messages* or commands are displayed in some more or less permanent form (e.g. printed on paper), rather than in a transient form (e.g. displayed on a screen of a *visual display* unit).

hard copy printer Any printing device which produces a print-out of information from a *computer* or *terminal* on paper, as distinct from a device which displays information on a screen as a transient record.

hard disc storage A data storage device used in *computers* to retain files of information, and in which *data* is recorded on tracks on both sides of a magnetized disc. The term 'hard disc' is used to distinguish such devices from *floppy discs* which work on the same principle but are made of less rigid material.

hardware A term used to define the physical components of a machine, particularly *computer* equipment, and devices containing logic elements used in *data processing* or communications machines. To be contrasted with *software*, which defines logical functions implemented as coding in a program.

HDLC ⇨ *high level data link control*.

header Information attached to a *message* or *packet* to provide for control of delivery to the appropriate *receiving station*, or for its management at the receiving station.

For example, in a *packet switching* network, a unit of application information is referred to as a *request/response unit*; it would be common to attach the following headers to it:

response header – identifies the packet to the *transmission control* and *data link control* (*DLC*) elements in the *network*;

transmission header – identifies the *routing* of the packet for *access path control* purposes;

link header and trailer – contains information relevant to checking and controlling transmission to overcome line errors; ⇨ *data link control*.

The precise titles and functions of headers may vary from one *protocol* to another, but the same principles apply.

header information Information attached to a *message* (or to a unit of *data* forming part of a message), and intended to impart information for use by the *communications control* features in a communications system. For example, header information may be used to: identify *destination addresses*, select *transmission paths*, determine *packet sequence*, provide *error checking*, and manage the *dialogue* between communicating *terminals*.

head-on collision An event which can arise in a two-way *data network* when two *nodes* secure the opposite end of a *data circuit* at the same instant.

hertz The definition of the frequency of a *signal* in cycles per second; e.g. a 50 kHz signal is an *AC signal* at 50,000 cycles per second.

The definition is named after Heinrich Hertz, a scientist who contributed to the discovery of the nature of *electromagnetic waves*.

hierarchic network A network in which some *nodes* exert more control than others; i.e. their *clocks* are more dominant than others.

Hi-Fi ⇨ *high fidelity*.

high definition television A television system in which a large number of lines are drawn to make a picture on a screen; e.g. more than the standard 625 lines recom-

mended by the *CCIR*. The amount of detail conveyed in a picture is increased by increasing the number of lines. However, increasing the number of lines also increases the requirements for *signal bandwidth*. The amount of detail which can be perceived by humans is another factor to be considered.

The expression *high definition display* is used in relation to TV as well as *visual display* units used in *text*, *data* and *graphical processing*. The term is not precise, and what may be considered high definition in one situation may not be so for another. ⇨ *visual acuity*.

high fidelity Relating to the processing of *sound signals* as an *electrical analog* of *sound waves*, and the reproduction of sound waves as a faithful representation of the original sound field. Systems which reproduce sound are never perfect, but those which can satisfy human requirements under controlled listening conditions are referred to as Hi-Fi or high fidelity systems.

high frequency (HF) A range of frequencies in the *band* from 3 to 30 MHz. ⇨ *frequency spectrum*.

high level control functions A concept related to *packet switching systems*, in which *control* is exercised by pairs of elements; one element of a pair is stationed at each end of the *transmission path*. The functions controlled by these elements include:

error-free transmission	– *high level data link control* (*HDLC*) element
addressing, routing, packetizing	– *access path control* element
dialogue management	– *data flow control* element

These so-called high level functions are performed by operating upon information conveyed as *headers* to *text* information, and are recorded and transmitted in the form of *bit patterns*. The use of such functions has been made possible by high-speed *digital transmission* facilities.

high level data link control (HDLC) A *communications control* standard developed by the International Standards Organization. The term also refers to a device and the associated communications *protocol* by means of which checking is undertaken to ensure that a transmission has taken place successfully. Under this standard, a *message* is transmitted as a series of *frames* which contain *bits* specifically added to the *message signal*. Checks are performed upon frames as they arrive at the *receiving terminal*. These checks ensure that all *information bits* are present, and include examination of redundant bits that have been added to frames at the *transmitting terminal*. Error frames are automatically detected and retransmitted.

Early forms of *data link control (DLC)* use different *characters* to represent *line control*, *device control* and *text*, but might use the same positions in a frame for such purposes. A disadvantage of this method is that corruption during transmission can convert a text character into a *control character*.

With modern high-speed communications protocols, line control information always occurs at a specified place in a frame, and distinctive *bit patterns* which have no relation to *alphabet* sets are used for control purposes. With this standard, line errors are much less likely to cause confusion, and greater efficiency arises in line utilization.

high level language A *programming language* used to give instructions to a *computer*, having a resemblance to a natural language such as English, or to mathematical notation. Compare with *low-level language*.

high pass filter A *filter* which allows all frequencies above a particular cut-off point to pass along a *circuit*, whilst blocking all frequencies below that point. Contrast with *low pass filter* and *band pass filter*.

high peak data rate The rate of traffic flow which arises when the *traffic* offered to a *network* approaches the maximum level planned in the design of the network.

high resolution A system for displaying information and graphic images is said to be of high resolution when fine details can be resolved upon a screen. There is no absolute definition of high resolution, but one might expect the number of *picture elements* making up the image to be of the order of a 500 × 500 matrix. ⇨ *resolution*.

high resolution display A *visual display* system capable of resolving a picture containing great detail. A relative expression for which no general standards can be given. ⇨ *resolution*.

high resolution graphics A standard for displaying *graphic information* which allows fine lines and details to be resolved. ⇨, for example, *alphageometric coding*.

high speed modem A device for connecting a *data terminal equipment* to a communications *line* and capable of operating at speeds up to 9,600 *bits* per second. ⇨ *modem*.

high-usage route A *transmission path* designed to take a high volume of *traffic* and which may be supported during periods of excessive traffic by auxiliary *routes*. ⇨ *fully provided route*.

highway A *transmission path* in which a number of *digital signals* may pass, but each separated in different *time slots*.

holding time The duration for which a particular *call* occupies a *channel*, or some part of a *telecommunication* system.

horizontal resolution A definition of the number of *picture elements* which can be resolved along a horizontal line drawn upon a display screen; e.g. in the British 625-line television system, 572 picture elements; in the USA 525-line system, 522 picture elements. ⇨ *resolution*.

horizontal tabulation (HT) A function represented as a special *format effector* in a *data communication code*, and serving as an instruction to move a *print mechanism* or the *cursor* of a *visual display* unit to the next predetermined position along a line of a document.

host processor A *computer* which runs *application programs* and exerts the primary level of control over activities in a *data network*. Until the 1970s, most data networks were dominated by the host processor which invariably appeared at the apex in a *tree network* hierarchy. Developments since then have tended to pass more control over communications activities to other *nodes* of the network, so that the host may be primarily noted for its control over *applications* and for its storage and processing capacity which are interfaced to the network by a *communications controller*.

hostile user A *user* who is intent on creating confusion or interruption of a *network*. Most networks, particularly public service networks, take measures to prevent unauthorized access to, or use of, the network.

housekeeping information Information recorded, transmitted and processed in a communication system to enable the system itself to perform its function in the handling of *user information*.

HT ⇨ *horizontal tabulation.*

hub polling A system of *polling* in which the *controller* activates the most distant *terminal* to enquire if it wishes to send or receive information, and that terminal responds accordingly before activating its neighbouring terminal directly, and so on until the polling cycle is complete.

hybrid A device in a speech transmission system, consisting of transformers that are arranged to convert a *two-wire channel* into a *four-wire channel*, and thus creates a separate wire pair for each direction of transmission. The separation of the two directions permits separate *amplification* of the *go* and *return paths*. Hybrids are also associated with *repeaters* used to amplify *message signals.* ⇨ *hybrid repeater.*

hybrid repeater In a speech transmission system, an arrangement of a *hybrid* and amplifying *repeaters*, which permits a *two-wire link* to be converted to a *four-wire link* for *amplification*, and then reconverted to a two-wire link.

hypergroup An assembly of 960 *speech channels* delivered over a single *co-axial tube* without loss of identity, using techniques of *frequency division multiplexing.* ⇨ *grouping.*

hypothetical reference circuit A *circuit* specification designated by the *CCITT* for reference purposes to enable design and operational issues to be expressed. There are many such recommendations, including those required for *television, sound, telephony, telegraphy* and *data networks*.

I

IA2 Abbreviation of *International Telegraph Alphabet No. 2.*, also known as *International Alphabet No. 2*.

IA5 ⇨ *International Alphabet No. 5*.

idle bytes Redundant units of *data* sent to a *line* when there are gaps in the sequence of data to be transmitted.

idle character A *character* inserted into a sequence of *message* characters to occupy particular *time slots* and to maintain timing; e.g. in a *time division multiplexor*.

idle state The condition that exists when a *line* or *terminal* is unused and awaiting a *call* to commence.

impulsive noise A form of interference arising in a *circuit* from electrical activity in adjacent equipment and which can be detected as a clicking noise in a radio or telephone *channel*.

in-band signalling Any form of *signalling* in which *control* signals to set up, progress, and *clear* a *call* are associated with the *channel* in which the *message signal* is transmitted and occur in a designated section of the *bandwidth*. Contrast with *out-band signalling*.

incoming buffers Storage space allocated to maintain a copy of incoming *message signals* until the particular *terminal*, *exchange* or *computer* is able to process the signals.

incoming call Any *message signal* being received by a *terminal*, or by an exchange, to be switched to an appropriate terminal *address* over *outgoing lines*.

incoming call rate The rate at which *calls* are received at a particular *terminal* or *node* in a system over a particular period.

incoming lines The communication *circuits* which are being used to bring *message signals* into an *exchange* or *node*, to be switched to appropriate *addresses* over *outgoing lines*.

incoming trunks The *trunk lines* connected to an *exchange* for the purpose of receiving incoming *traffic* from another exchange.

inductance A property which links the magnetic flux in a *circuit* to the current flowing in the circuit or in an adjacent *conductor*. This phenomenon has to be considered in the design of circuits used for transmission of *message signals*.

information A non-specific term meaning any *signals* transmitted or stored in a *network*. The term should be qualified; e.g. *user* information implies information required by the users of a system rather than *housekeeping information*, generated to allow the communication system to function correctly.

information bearer channel A *channel* provided for communication in a *data network* which is able to carry all information to allow communication, including, in addition to the *user data*, certain *control information* and data synchronizing signals. The information bearer channel may, therefore, operate at a *data signalling rate* higher than that required solely for transmission of user data.

information bits This term is frequently used to denote *user* information in *binary coded* form, as distinct from *synchronization* bits or *check bits*, which may be added to allow a system to handle the user information correctly.

information content That part of a *character* or *message* which contains pulses representing *user* information rather than *signalling* or *addressing* information required by the *network*.

information feedback system A system in which the *receiving station* always returns *blocks* of *data* to the *transmitting station* to allow the *transmitting terminal* to check whether there has been an erroneous transmission. Blocks designated as errors are retransmitted.

information provider (IP) The generic title given to persons or organizations who supply information to be made available to *users* in a *videotex* service.

information rate A term referring to the speed with which information is transferred over a *circuit*.

Synonymous with *data transfer rate* and contrasted with *data signalling rate* and *modulation rate*.

information retrieval Pertaining to a *data processing system* specifically designed to provide search and retrieval facilities in response to random requests by *users* for responses from a *database*.

information security Relating to the extent to which procedures are established to prevent purposeful or accidental access to information held in a system, and to prevent disclosure, modification or destruction of *user* information.

information separators Special symbols forming part of a *data communications code* and used to separate *fields* of *information* as required by an *application*.

Examples include: *FS* (*file separator*), *GS* (*group separator*), *RS* (*record separator*) and *US* (*unit separator*).

information theory A body of mathematical theory relating to the factors which affect the transmission of information over a set of network facilities. The theory concerns itself with the channel capacity, noise, the information content, and other factors affecting information transmission. ⇨ *Shannons Law*.

information transfer The end result that occurs in sending information from one *terminal* to another over a *network*.

infra-red light Radiation beyond the *visible light spectrum* covering a range from about 730 nanometres to about one millimetre in *wavelength*. ⇨ *electromagnetic radiation*, and ⇨ Appendix 5.

initial signal unit The first *signal* of any *signal message* which uses more than one *signal unit*.

in-plant communication Relating to the use of *private networks* for communication within a building or group of buildings without utilization of facilities and *circuits* provided by a *common carrier*.

in-plant equipment A term used to describe the internal facilities for communication within an organization. Such equipment may serve internal communication functions, without access to *common carrier* facilities provided by the public service authority, or may connect to common carrier facilities for external communications.

input channel Any *communication channel* which provides input *signals* to a device or process.

input data signalling rate The rate at which a device can receive information transmitted to it, and usually expressed in *bits*-per-second.

insertion loss The overall *loss* of power that arises in a *circuit* attributable to the losses introduced by each and every circuit component. The insertion loss is usually measured in *Bels* or *deciBels* end to end, using a test frequency of 800 Hz. Also known as *overall loss*.

in-slot signalling *Signalling* information

associated with a *channel* and transmitted at a predetermined position within the *time slot* allocated to the channel.

instantaneous traffic The average number of concurrent *calls* in progress in a given system is known as the instantaneous traffic and is expressed in *erlangs*; e.g. 24 concurrent calls is 24 erlangs of *traffic*.

instantaneous traffic level The number of *calls* in progress at a particular moment in time. ⇨ *traffic volume* and *instantaneous traffic*.

integrated digital exchange An *exchange* using digital technology and in which all *traffic* is received and distributed as *digital traffic*. *Voice signals* are converted from *analog* to digital form using *pulse code modulation* (*PCM*) techniques, and are transmitted using *time division multiplexing* (*TDM*) techniques. Exchanges using this technology are said to be integrated because it is possible to handle all basic services (e.g. *telephone*, *data*, *telex*, *videotex* and *facsimile*) using the same exchanges and *trunk network*.

integrated digital transmission and switching system A *network* in which connections are made by using *digital switches*. With this type of system, speech, *data* and other *signals* are transmitted as *digital signals*, using the techniques of *time division multiplexing* in both the transmission and the *switching equipment*. In such a system, *speech signals* are converted to digital signals by analog to digital converters at the *transmitting* and *receiving stations*. Speech is then transmitted digitally over a *four-wire circuit* and the *transmission loss* is independent of the distance and the number of digital exchanges used to carry the *call*. ⇨ general article on *switching equipment*.

intelligent controller A device used to control a *cluster* of remote *terminals*, handling their *control* requirements for communication with a *network* or central *node*

in a system. Such a device may be programmable to carry out specific *application* tasks required by *users*.

intelligent terminal A *terminal* which has storage capacity and processing power, thus enabling complex logical functions to be performed within the terminal in support of the *users application*, or to accommodate *high-level control functions* required by the *interface* requirements of the *network*. Contrast with *dumb terminal*.

intelligent time division multiplexor A device which acts as a *time division multiplexor* (*TDM*) but contains a *microprocessor* control device which allocates the available *bandwidth* (i.e. *time slots*) dynamically, to improve the utilization of the *channel*. With more conventional TDM devices, the time slots are allocated whether a particular channel is active or not, and *idle characters* are inserted if the channel is not activated. Also known as *statistical multiplexor*.

intelligible crosstalk *Crosstalk* which results in intelligible *speech signals* being transferred from one *circuit* to intrude on another adjacent circuit.

interactive applications A *data processing system* in which the *users* are directly connected, when required, to the *computer*, using a *terminal* and communication lines; and in which the computer program respoi ds to events initiated by users to maintain an accurate up-to-date record of the events or objects being controlled.

interchange circuit A *circuit* between two devices over which *signals* are exchanged to allow communication to take place; e.g. between a *data terminal* (*DTE*) and a *modem*. ⇨ *communications interface*.

interchange signals *Signals* passed between two devices, or systems, which have separate functions. For example, between a *modem* and its associated *data terminal*

(*DTE*) to effect *signalling* to a distant terminal. Examples are described under *request to send*, *ready for sending*, *carrier detector*, *calling indicator* and *data terminal ready*. ⇨ *communications interface*.

interchange specifications A specification defining the electrical and logical connection between any two devices; e.g. between a *terminal* and a *modem*.

intercom A simple system to provide two-way communication between persons in different rooms. Each person is usually equipped with a *microphone* and a *loudspeaker* and these are connected to the equipment at the remote location by a cable. *Voice signals* are amplified to pass from one location to the other.

inter-exchange signals The signals that are passed between *telephone exchanges* (*offices*), in setting up, maintaining and clearing a trunk *call*. ⇨ *signalling*.

interface A specification of the rules by which interaction between two separate functional units can be made to operate to conform with overall system requirements. Since most modern systems are modular in nature, with perhaps several hundred units in any one overall system, there are many levels of interface to be considered. An interface specification may include logical, electrical and mechanical specifications. Many standard interface specifications exist for different classes of system or device, and international organizations such as the *CCITT* are responsible for producing standards for international compatibility.

interface compatibility When introducing a new item of equipment into an existing system of communication, it is necessary to ensure that its electrical and mechanical characteristics *interface* with existing components. ⇨ *functional compatibility*.

interlaced scanning A technique used in television broadcasting to reduce the *signal*

bandwidth by limiting the *picture frequency* and arranging for it to be a sub-multiple of the *field frequency*. Usually there are two *fields* per complete picture; thus a picture of, say, 500 *lines* can be achieved by interlacing two field sequences of 250 lines. The interlaced sequences are arranged so that one traces the odd number lines 1, 3, 5 ..., and the other the even lines 2, 4, 6, etc.

The picture frequency under this arrangement is half the field frequency and the *flicker effect* is observed to affect small areas of the screen rather than the total picture.

inter-layer interface A precise definition of the procedures and logical structures providing for interaction between the different layers of *control* in a *network architecture*. An example of such an architecture is provided by the *ISO reference model for open systems architecture*.

intermediate exchange A *transit exchange* at a point in a *network* and part of a connection for a specific *call*. Not the *first exchange* or the *last exchange* (q.v.).

International Alphabet No. 2 (IA2) A *data communications code* used primarily for *telex* communication in which up to 52 *alphanumeric* characters and symbols can be represented plus *carriage return*, *line feed*, and *figures* and *letters shift* characters.

International Alphabet No. 5 (IA5) This is a *data communications code* developed as an international standard to allow telegraphic and *data transmission*. It originated from a standard, formulated by the American Standards Association, known as *ASCII* (American Standard Code for Information Interchange). The standard was further developed by the *International Organization for Standards* (*ISO*) and the *CCITT*, and was ratified by them in 1968.

The code is shown in Appendix 1.

international circuit Any *circuit* between

international exchanges situated in different countries.

international dialling prefix The set of *digits* which has to be dialled by a *user* wishing to make an international call, and used to obtain access to the outgoing *international exchange*.

international exchange An *exchange* which provides connection for *traffic* on to *international circuits*.

international gateway An *exchange* used to switch *traffic* (*voice*, *telex* or *data*) between a national communications *network* and an international network.

international leased circuit The complete *transmission path* from a *terminal* in one country to a terminal in another country and reserved exclusively for the use of a particular organization or person.

international number A number which excludes the *international prefix* providing access to the international *network*, but which includes the *country code* and the national subscriber number for the *subscriber* concerned.

International Organization for Standards (ISO) An international body, including standards groups from many countries, which develops standards for goods and services to facilitate international trade and exchange. Particular standards for *data communication* are developed by ISO Technical Committee 97.

international prefix A combination of *digits* which has to be dialled by any *subscriber* to obtain access to the international *network*, and is then followed by the full *international number* of the subscriber to be called.

international sound programme centre A centre in which *audio circuits* for *sound programmes* terminate and in which connections can be made to other sound *circuits* to allow for supervision and distribution of sound transmission between different countries.

international switching centre An *exchange* for switching *traffic* between different countries over *international circuits*.

International Telecommunication Union (ITU) ⇨ *Union Internationale des Télécommunications*.

International Telegraph Alphabet No. 2 (IA2) A *data communication code* used primarily for *message* systems using the *teletypewriter* and the international *telegraph network*. It is a *5-bit code* which uses *shift characters* to permit up to 64 characters to be represented. The code was first ratified by the *ITU* in 1932.

International Telegraph Convention An international body established under the auspices of the *ITU* to harmonize standards in telegraphic communication.

international television centre A centre in which *television circuits* for carrying *video signals* terminate and in which connections can be made to other television circuits, to allow for supervision and distribution of television programmes between different countries.

international transit exchange A *telephone exchange* situated in one country but intended as a centre for switching *calls* between other countries.

international trunk dialling The method for directly calling a telephone *subscriber* in another country, in which the *calling party* adds a *prefix* to the basic number of the *called party*. This prefix consists of a special code which connects the caller to the *international gateway* plus a *country code*. Standard country codes have been created by the *CCITT*.

inter-office signals The *signals* that are passed between *telephone exchanges* in setting up, maintaining and clearing a trunk *call*.

inter-office trunk ⇨ *junction circuit.*

interrupted isochronous transmission The transmission of *synchronous data* in bursts over a *public data network* where the rate of transmission on the *information bearer channel* is higher than the *input signalling rate* of the receiving device. Also known as *burst isochronous transmission.*

interval A period of time selected for a specified purpose.

interworking The process by which two systems can interact. For example, a *terminal* in one *data network* can interact with one in another network. Or the means by which dissimilar terminals can communicate through a network.

intranode addressing In a *data network*, the *addressing information* concerned with *users* connected to the same *node*. The logical equivalent to a *local number* in a *telephone network*.

intranode routing The operations concerned with providing an *access path* for *users* connected to the same *node* in a *data network*.

ionosphere A layer of air, surrounding the planet Earth, which is ionized; i.e. it consists of charged gases which provide conducting layers. The layers vary in thickness from 50 to 400 kilometres above the earth, and they have different effects upon radio broadcasting in different *bands* of the *radio frequency spectrum*. At very low frequencies, *radio waves* travel around the earth as though in a *waveguide* between the earth and the ionosphere. However, at higher frequencies, waves are reflected from the ionosphere and are sometimes referred to as *sky waves*. Both *medium frequency* and *high frequency* broadcasting makes use of sky waves. Transmissions in these bands are affected by variations in the ionosphere and by overcrowding of *broadcasts*, as well as atmospheric conditions such as are caused by sun spots. Above the high frequency band, *radio signals* are not affected by the ionosphere.

ISO Abbreviation of *International Standards Organization.*

ISO reference model for open systems architecture This model for the construction of *data networks* was first published by the *International Organization for Standards (ISO)* in November 1978. It was intended to provide a basis for *open systems architecture*; i.e. a network in which there is no single central control point, but one in which *control* resides in the various *nodes* of the network operating to common standards. These standards are expressed in an *architecture* which provides seven control levels.

Each control level provides a clear definition of the *protocols* and formats and allows a *peer interaction* between *users* who have implemented the architecture; i.e. the control levels of one user communicate with the equivalent level of another user. Thus, two users who have correctly implemented only the first two levels of control are able to exchange *information* over a network.

The control levels have been defined as follows:

LEVEL 1 – this is called the *physical control layer* and defines the *physical interface* between *data terminal equipment (DTE)* and *data circuit terminating equipment (DCE)*; it presents the electrical characteristics and *signalling* needed to establish, maintain and clear the physical connection between line terminating equipment.

LEVEL 2 – this is known as the *link control* and equates with the *high level data link control (HDLC)*. It provides the protocol for checking that information transmitted in small units known as *frames* has been correctly exchanged across the network between two *terminals*.

LEVEL 3 – this is called the *network con-*

trol layer, and provides control between two adjacent network nodes, and between *DTE* and the network; e.g. the *X25 packet network* standard which allows packet formatting and *virtual circuits*.

(The standards for these first three levels are defined; the detailed standards for the following levels are under discussion at the time of writing.)

LEVEL 4 – known as the *end-to-end layer*. This provides control from *user node* to user node including *addressing*, *data assurance* and *flow control*.

LEVEL 5 – this is known as the *session control layer* and provides for establishing, maintaining and releasing logical connections for *data transfer*.

LEVEL 6 – this layer is known as *presentation control* and provides for data formats and transformations; e.g. those required for *visual display* screens or printers.

LEVEL 7 – this is the *application layer*.

The objective of the ISO in publishing the model is to achieve progress towards a complete family of standards for operating on *public data networks*. Implementations of the standard have been achieved; for example, the X25 standard defines a

three-level protocol corresponding to the first three levels of the ISO model. It is expected that convergence towards a single standard covering all layers of the ISO model will take several years, but the goal of having any terminal being able to access any application is being brought nearer.

isochronous system A communication system in which timing information is transmitted on the *channel*, as well as *data*, to establish a common *time interval* for *transmitting* and *receiving stations*. More commonly referred to as *synchronous system*. Contrast with *anisochronous system*.

ISPC ⇨ *international sound programme centre*.

ITC ⇨ *international television centre*.

ITDM ⇨ *intelligent time division multiplexor*.

ITU Abbreviation of *International Telecommunication Union*, also known as *Union Internationale des Télécommunications* (*UIT*). Refer to this entry for description of UIT objectives.

J

jitter A signal impairment causing *bits* of a *digital signal* to be advanced or retarded, relative to the *time slots* allocated. A timing fault of this nature would probably cause errors at the *receiving terminal* or *node*.

judder Related to *facsimile* and describing a condition in which there is a lack of uniformity in the scanning of a picture, resulting in overlapping of elements of the picture.

junction circuit A *circuit* provided between two *local exchanges* in an *exchange hierarchy*. Such a circuit acts as a *trunk circuit* by connecting two local exchanges without *routing* via a *trunk exchange*. In the USA it is referred to as an *inter-office trunk*. In Britain the term is also used to describe the link between a local exchange and a trunk exchange.

justification A process concerned with altering the *bit rate* of a *digital signal* on a *channel* so that it is brought to conformity with another rate in a controlled manner. A technique used in *time division multiplexing*. Also known as *bit stuffing* and *digital filling*.

justifying digit A digit inserted into a *digit time slot* as part of the process of *justification* and distinguished within the process as not an *information bit*.

K

k Abbreviation for kilo, meaning a thousand, as in *kilocycles* (*k*) or *kilobits* per second (kbs).

kell factor A factor used in dealing with human perception of images in a *visual display* system; e.g. television. This factor has been derived from statistical tests to determine the ability of humans to resolve alternate black and white elements. There are a range of kell factors suggested by experiments, but a factor of 0·7 is used in television systems. Thus, if it is accepted that the average television viewer can resolve 425 lines (see *visual acuity*), then 607 lines are required to produce this *resolution*; 607 × 0·7 gives 425 lines.

key A term used mostly in *data processing* to indicate a *data element* which identifies a particular information record; e.g. a stock item number on a file containing inventory details.

keyboard A device for entering information in a *data terminal*. For example, allowing for all the functions and *characters* of the *International Alphabet No. 5* to be controlled by the human operator of a *terminal*.

keyboard send and receive (KSR) Pertaining to a *terminal* used as a *data printer* and also having a *keyboard* to enter *alphanumeric* information. It can be used as a *remote printer* or a *data entry terminal*.

keypad A device providing a limited set of *characters* and functions to be entered into a *data terminal*. ⇨, for example, *videotex terminal*.

key-to-tape machine A device which captures information to be transmitted at a later time. It consists of a *keyboard* containing keys for *alphanumeric* characters. *Data* entered through the keyboard is encoded on a storage medium such as *paper tape* or *magnetic tape*.

kilobits A unit of *data* volume, i.e. one thousand *binary digits*. For example, a particular *channel* may be said to handle 9·6 kilobits per second.

kilocycle One thousand cycles per second.

KSR Abbreviation of *keyboard send and receive*. A type of *data printer*, used for *remote printing* or as a *data entry terminal*.

L

label An element of information attached to a *message* for some identification purpose; e.g. to identify a *circuit* with which the message is associated.

LAN Abbreviation of *local area network*.

LAP Abbreviation of *link access procedure*.

LAPB Abbreviation of *link access procedure B*.

large scale integration (LSI) The application of *micro-electronic technology* to the integration of hundreds or thousands of components on to a single *silicon chip* of small physical size. The design, manufacture and testing of such chips can be aided by automated techniques.

laser A device used to generate light in a coherent manner, e.g. a beam of light interrupted or phased to represent digitally encoded data. The term laser is derived as an abbreviation of Light Amplification by Stimulated Emission of Radiation.

Lasers are used to generate light signals for transmission over fiber optic systems; ⇨ general article entitled *optical fibers*.

last exchange Refers to the *exchange* closest to a *called party* or to the last exchange in any connection which utilizes a specific *signalling* system; e.g. the last *No. 6 exchange*. ⇨ *first exchange* and *intermediate exchange*.

layered architecture A *network architecture* in which the various levels of *control* are logically separated from one another and from the physical details of the communication medium. The *ISO* has produced a standard which defines a hierarchy of 7 levels of *protocol*. This is an example of a layered architecture. The virtue of implementing systems in accordance with standards of this nature is that *users* requiring more simple functions can implement the lower levels only, and those requiring increasing levels of complexity can progressively implement the higher levels. This also allows for the practical growth of facilities at successive stages of *network implementation*. ⇨ *ISO reference model for open systems architecture*.

LE Abbreviation of *local exchange*.

leased circuit A *circuit* hired by a particular person or organization to provide a connection between two locations, the circuit not being available for use by another party.

leased lines *Transmission channels* for *voice* or *data* which are used exclusively for a particular organization for communication between predetermined locations. The *lines* are hired from the telephone company or transmission authority to provide transmission facilities where the requirements are such that dependence upon a *public switched network* is uneconomic or not sufficiently reliable.

least significant bit In any *binary number*, or unit of *binary coded information*, the lowest order *binary digit* (i.e. 2^0 position). In a communications system this may not necessarily be the first *bit* transmitted.

LED Abbreviation of *light emitting diode*, which is a device used to display operating conditions and status on the fascia panels of communications devices.

letters shift A special *character* in a *telegraph code* or *data communication code* which has the function of designating all subsequent codes in a transmission sequence as being letters (A to Z). The letters shift remains in force until a *figures shift* occurs.

LF ⇨ *line feed*.

light emitting diode A device often used to provide a *visual display* on the fascia panel of communications devices.

light frequency Light is a form of *electromagnetic radiation*. *Infra-red light*, in the frequency from 1 THz to 100 THz, is today being considered for transmission in *optical waveguides*. 1 THz (Tera) is equal to 1,000,000,000,000 cycles per second. ⇨ *electromagnetic waves*.

light waves *Electromagnetic radiation* in the light spectrum which travels in space. With the introduction of *optical waveguides*, light waves are being increasingly used to carry communications *traffic*.

limit test
1. A test made to check the sensitivity of components in an equipment, in an attempt to isolate a possible cause of intermittent failure or poor service. For example, creating a temporary reduction in operating voltage.
2. Any test made to ascertain if any measurable entity would appear outside of predetermined limits.

limited distance modem A simplified *modem* used over short distances or on *private lines* where it is not necessary to comply with the more stringent requirements for a modem approved for *interfacing* to the *public switched telephone network*. Also known as a *data service unit*.

line Any *circuit* providing a connection between two points. It may consist of two wires only, or of four wires. In *telephony*, a two-wire *line* is used to connect a *telephone* to a *local exchange*, but four-wire lines are used in the *network* to permit line *amplification* and *multiplexing* to take place.

line access point A physical point for connection of a *terminal* to a *line*.

line adaptor A simplified form of *modem* which is used for connecting a *data terminal* (*DTE*) to a *private network* or certain *public data networks*. Not generally acceptable for connection to the *public switched telephone network*. ⇨ *data service unit*.

line code A *code* used for transmission purposes which may use pulses chosen to represent *characters* which differ from the representation used in *data terminals* (*DTE*).

line control Pertaining to the discipline and *signalling* method which enables *terminal* devices connected to a *transmission path* to communicate with one another.

line driver A simplified *modem* for connecting a *data terminal* (*DTE*) directly to a *line*, and used for connection to *private networks* or certain *public data networks*. Not usually suitable where transmission is necessary using the *public switched telephone network*. Also known as a *data service unit*.

line feed (LF) A function represented as a special *format effector* in a *data communication code*, and serving as an instruction to move a *print mechanism* or the *cursor* of a *visual display* unit to the beginning of the next *line*.

line flyback An interval during the scanning operation of a television picture, in which the *electron beam* is blocked off and the *raster* is repositioned to begin a new *line scan*. ⇨ *video signal*.

line frequency The frequency of *line scanning* operations in the generation of a picture in a television system. The frequency is given by the number of *lines* × the *picture frequency*, and is measured in cycles per second (Hz); typical values are: USA 525-line system, 15,750 Hz; and *CCIR* 625-line system, 15,625 Hz.

line isolator A device which connects a *terminal* to a communications *line* to protect the *network* from high voltages or extraneous *signal* frequencies inside the terminal.

line link This term describes a *transmission path* of uniform *bandwidth* between any nominated points including all *lines* and equipment necessary to maintain the path.

line loop test A method of testing a *line* and its *modem* using test equipment. ⇨ *loop back test*.

line-out-of-service signal A *signal* sent on a *backward channel* to denote that the *line* to the *called party* is not serviceable.

line period The time taken in the creation of a television picture to trace the *raster* movement, equivalent to a single horizontal *line scan* and the corresponding *line flyback*. This entails a movement of the *electron beam* across the screen and back, including a downward movement ready to start the next forward stroke. For the *CCIR* 625-line system, the line period is 64 microseconds. ⇨ *video signal*.

line scan The horizontal movement of the *raster* during the scanning operation to create a television picture. ⇨ *video signal*.

line-up period The duration of time required to prepare a *television circuit* by a communications authority before handing it over for operational use by a broadcasting authority.

link The complete assembly of circuit sections which make up a *transmission path* for a particular purpose.

link access procedure ((LAP) and (LAPB)) A procedure which operates in a *packet switching network* for data interchange, and is responsible for functions involving framing, *synchronization* control, *error detection*, and for connections between the *terminal* and a *network node*.

LAPB is a standard version of this procedure recommended for the *CCITT X25* and uses a subset of the *ISO* HDLC (the *high level data link control* specified by the *International Organization for Standards*).

Link control procedures operate at level 2 of the seven levels provisionally defined by ISO for *open systems architecture* – ISO TC97XC16N117 (November 1978).

link control In managing the communication between two *terminals* in a *data network*, certain procedures are operated to ensure that the *data* received is an error-free replica of the data transmitted. These procedures, known as *link control*, vary from one standard to another, but usually employ the addition of redundant *bits* to *frames* of *information* which are automatically checked. Automatic retransmission is requested if errors are detected at the *receiving station*. Common standards include *high level data link control* (*HDLC*) developed by the International Standards Organization, and *advanced data communication control protocol* (*ADCCP*) developed by the American National Standards Institute. Also known as *data link control* (*DLC*).

link header and link trailer In *packet switching networks* the *data link control* (*DLC*) adds *information bits* to the beginning and end of each *packet* to assist in the *error checking* procedures which are performed to ensure that *data* is not lost or corrupted. ⇨ *link control*.

link level The procedures and functions in *data communication* concerned with *link control*; i.e. in managing the error-free transmission and reception of *data*. It relates to level 2 of the *ISO reference model for open systems architecture*.

link management The procedures in a communications system concerned with allocation of *channels* and the *resolution* of *contention* for the available channels.

listening tests Tests conducted by a telephone authority to establish the preference of *users* in regard to the loudness of the *message signal.* Also used to test the sensitivity of human users to *bandwidth* and *signal-to-noise ratio.* Such tests do not depend upon assumptions about the theoretical qualities of the human ear, but require a large number of tests to be conducted to compile responses, known as *perception data.* From these, statistical conclusions are derived.

loading A technique adopted to reduce the power loss which occurs in *circuits* due to the inherent *capacitance* between parallel wires in close proximity. Loading entails introducing coils which provide *inductance* to counteract capacitance.

local area network A *network* designed to provide facilities for user communication within a defined building or plant and which does not necessarily use public service facilities or standards. An example of such a network is *Ethernet,* a proprietary system intended for high-speed *traffic* between *users* at distances up to 2·5 kilometres.

local call rate The *tariff* charged to a *user* of a *telephone network* for a *call* within a local area. Usually a cheap charge (or sometimes no charge), compared to the tariff for a *trunk* call.

local circuit
1. In *telephony,* a *circuit* between *subscribers* and a *local exchange.*
2. In *data communication,* a circuit forming part of an *in-plant* network and not one provided by a communications *carrier* such as a *PTT.*

local exchange An *exchange* to which a population of *terminals* in a particular geographic area are directly connected by means of *local lines.* The exchange itself has access to other exchanges and to national *trunk circuits.*

local lines The *lines* connecting a *terminal*

to a *local exchange,* and providing a *transmission channel* by which the terminal is able to connect with an *exchange hierarchy* forming an overall *network.*

local loop A *local area network* providing communication over a shared *broad bandwidth* channel to which several hundred *users* may be attached. Mainly used for high-speed *in-plant communication* (an example is described under the entry for *Ethernet*).

local mode Certain communications devices can operate to provide useful functions when not connected to a *communications channel.* In this condition they are said to operate in local mode. For example, tests can be run on the local equipment, or *data processing* operations such as collection and validation of *data* can be performed.

local network
1. In *telephony,* that part of a *telephone network* which embraces the subscriber telephones, the *lines* to *local exchanges* and the local exchanges themselves. The term distinguishes these facilities from the *trunk network* which includes the *trunk exchanges* and *trunk lines.*
2. The term is also applied in *data communication* to refer to *in-plant* networks which do not depend upon *public switched* or *leased circuits* to provide *transmission paths.* Also known as *local area network.*

local number A *number* identifying a *subscriber* on the same *local exchange* as another subscriber, or an abbreviated *telex* number used for local *traffic* in a telex network.

local, own exchange, call A telephone *call* which is between two *users* whose *telephones* are connected to the same *local exchange.* The call thus utilizes *local lines* but does not use *trunk circuits* which provide *transmission paths* between distant *trunk exchanges.*

local signals The *signals* that are passed between a *telephone* and the *local exchange* in setting up, maintaining and clearing a call. ⇨ *signalling.*

local telephone network Pertaining to the *network* involving the *local circuits* between *subscribers* and a *telephone exchange*, but not including *trunk circuits* which connect *trunk exchanges.*

lockout facilities Facilities, usually implemented by *software*, which prevent interference between different *users* who may be simultaneously trying to update the same records in a *database*, or to make reservations upon the same objects. The system would automatically assign priority according to the sequence in which user requests are received, and the possible *contention* is not apparent to the users.

log-on The act performed by a *user* when he enters a communication system to conduct a particular *session*, usually involving procedures to enable the system to check the authority of the particular user to perform allowable functions.

log-off The act performed by a *user* when he concludes a particular *session* in a communication system, for which the time being terminates operations involving the use of system functions by that user.

logical channel There are various techniques of *multiplexing*, which enable several *data channels* to be available using a single physical *transmission link*. The concept of a logical channel is used to identify each data channel within the *network*, and to distinguish it from other channels sharing the same physical resources. ⇨ *logical channel number.*

logical channel number In any form of *multiplexing* it is possible to transmit hundreds or thousands of *calls* simultaneously over the same *transmission path*. Each separate call is transmitted along a *channel* which occupies a particular frequency or a particular *time slot*. These channels are identified in the *communication control* system by unique numbers known as logical channel numbers. ⇨ *frequency division multiplexing* and *time division multiplexing.*

logical interface The rules governing the way in which two devices must interact, including the identification of *signals* passing between the devices and the responses given by one to another under given conditions. Compare with *electrical interface* and *mechanical interface.*

lone signal unit A *signal unit* consisting of only one unit of information, rather than a *message* consisting of several signal units.

long circuit A term used in *telephony* to define a *circuit* which requires *echo suppression.*

long-term store A storage medium used to hold *data* which must be retained for a long time. For example, in a *message switching system*, to retain *messages* until they can be sent to the correct *destination address*. Long term store is usually a magnetic disc forming part of a *computer* system.

long wave Radio broadcasts in the *low frequency* band; i.e. 30 to 300 kHz.

longitudinal judder A condition occurring in *facsimile* devices in which a picture is reproduced with irregular lines due to inaccurate rotation of a scanning device.

longitudinal redundancy checking (LRC) One of the simplest forms of *redundancy checking* in which a *check character* is added to each *block* of transmitted *data*. The check character is computed to make each row of *bits* formed by the same bit positions in each successive character into a specific parity; e.g. an even number of 1-bits equals *even parity*. Thus, using 8-bit characters, each block would consist of 8 strings of bits, the number of bits in each string being dependent on the block length.

The parity of each string is checked at the *receiving station*, and, if parity errors are detected, the block is classed as an error and retransmission is requested. When combined with *parity checking* on each character position, LRC can achieve a powerful method of *error detection*. ⇨ *cyclic redundancy checking*.

loop back test A test of the quality and performance of a *line*, and/or its terminating equipment, achieved by looping the outgoing and incoming paths to an item of test equipment used to create operational conditions. The test equipment will generate test patterns which are analysed on receipt over the *return path*.

loop-disconnect signalling A traditional method of *signalling* in a *telephone system*, in which pulses representing the called *address* are generated by disconnection of a *circuit*. A dial or push-button may be used for the purpose.

loop network A *network* in which the various *nodes* are interconnected along a *transmission link* represented as the circumference of a wheel. The nodes appear as points along the circumference, and communication between two nodes must proceed via any intermediate nodes along the loop. Also known as *ring network*. ⇨ *network topologies*.

loop test ⇨ *loop back test*.

loss The loss of signal power caused by the characteristics of the cable, including energy consumed in overcoming resistance in the cable, and the effects of *attenuation*.

loss/frequency response A term used to describe the variation in *loss* of power which can be experienced at different parts of the *bandwidth* of a given *circuit*. The figures quoted to describe this will be in *decibels* and be expressed relative to the loss experienced for a given frequency; e.g. the *insertion loss* at 800 Hz. This phenomenon

is not generally considered important in *speech communication* over a *telephone channel* but may be significant in *data communication*.

loss-of-frame alignment An error condition arising in *pulse code modulation* systems in which a receiving device is unable to determine the correct positioning of *frames* in the incoming *signal*.

lost call A *call* which cannot be accepted by a *network* due to congestion of the available *circuits*.

loudspeaker A device which generates *sound waves* from *electrical analog* signals representing sound. A *transducer* which converts electrical energy into sound energy.

low delay A quality desired in a *network* in which there should be as little delay as possible in the delivery of a *frame* of *information* at any planned level of *offered traffic*.

low frequency (LF) *Radio waves* in the *frequency band* of 30 to 300 kHz, sometimes known as the *long wave*

low level language A *programming language* used to give instructions to a *computer* but having a form and structure convenient for representation within the machine, and not designed to have a form easily understood by humans. Compare with *high level language*.

low pass filter A *filter* used to allow all frequencies below a particular cut-off point to pass along a *circuit* without *attenuation* whilst blocking all frequencies above the point. Contrast with *high pass filter* and *band pass filter*.

low resolution Pertaining to a *visual display* system, and defining one which cannot be used to present *graphic information* in fine detail. An example would be the *alphamosaic graphics* displays used in early *videotex systems.* Contrasted with *high resolution*, but ⇨ *resolution*.

low speed modem A device for connecting a *data terminal* equipment to a communications *line* and capable of operating at speeds up to 1200 *bits* per second. ⇨ *modem*.

lower sideband The process known as *modulation* (imprinting a *message signal* on to a *carrier wave*) generates a series of harmonic frequencies known as the *spectrum envelope*. The principal outputs generated are two *sideband* signals equally displaced in frequency about the *carrier signal*. One, the *upper sideband*, is above the frequency of the carrier; the other is below and is known as the lower sideband. ⇨ *modulation*.

LRC ⇨ *longitudinal redundancy checking*.

LSB Abbreviation of *least significant bit*.

LSI Abbreviation of *large scale integration*. ⇨ *micro-electronic technology*.

luminance A measure of the brightness of a light source, usually expressed in candelas per square metre.

luminous flux The total *photometric power* generated by a source in all directions, where *photometry* relates to the measurement of light energy to account for the human perception of such energy which varies by sensitivity to different *wavelengths*. ⇨ *visibility function*.

M

magnetic deflection coils Devices used to create movement of a *scanning beam* inside a television camera or *television receiver*. In a *cathode ray tube*, they are positioned around the path from the *electron gun* to the screen. They are activated to deflect the *electron beam* across and down the screen to trace out the *raster* movement required in a television system.

magnetic tape A medium for storing *data* in *computers* in which *digital signals* are encoded on to a magnetizable strip of material. Once information has been recorded, it can be retrieved by reading the tape serially from the beginning – thus, information is stored on magnetic tape when it is to be archived for a time, and it is not expected to be immediately accessible to *users*.

main frame A *computer* forming the central part of a *data processing system* and usually a general-purpose computer able to run a mixed and heavy workload including *on-line systems*, *batch systems*, *program development*, and *database management*.

It is common to connect a *main frame computer* to a *network* by a *front end processor* or *communications processor*; thus, the communications aspects are handled by a specialized processor, while the main frame runs the *applications* task.

A main frame is usually much more powerful than a *minicomputer* and the term is more commonly used to distinguish large general-purpose machines from minicomputers and *microcomputers*.

main frame computer ⇨ *main frame*.

main memory That part of a *computer* store into which programs and *data* must be transferred before processing can take place. The main memory consists of thousands of storage locations which can be directly addressed by the central processing unit with very short access times. Contrast with *backing storage* which contains programs and data stored on a medium such as a magnetic disc. The access to main memory locations is measured in nanoseconds, whereas backing storage access is measured in milliseconds.

main network ⇨ *trunk network*.

main trunk A communication link providing a major route between *exchanges* in a *trunk network*. Also known as *trunk circuit*.

maintenance The operations concerned with setting up, monitoring, identifying faults, and repairing *hardware*, *software*, *circuits* and equipment within the limits intended to provide a service to a prescribed level of quality.

maintenance control system A subsystem in a *digital exchange* which diagnoses faults in the system and provides information to assist engineers to isolate and correct faults.

man/machine interface
1. A subsystem of a communications *network* which provides for interaction between the command centre of the network and human operators responsible for *network management*. The commands issued across the *interface* allow for monitoring, controlling and maintaining the network and its component parts.
2. Any procedure or *protocol* which allows a human operator to access and operate a system.

management signals *Signals* concerned with the maintenance or operational management of a *network*.

management statistics A set of *information*, usually gathered automatically by a system, to provide basic *data* needed for short- and long-term planning for the operation and enhancement of a *network*.

manual answer A facility in which a *call* can be established only if a human operator is in attendance, to complete a manual operation to signify that the call will be accepted.

manual answering Pertaining to any system in which a human operator must be present to perform an operation in order to permit an *incoming call* to be accepted.

manual calling A system in which the *address* of a *called terminal* is entered by a human operator (and at an undefined rate), but the address *characters* themselves may be generated by a *data terminal* as a result of the operator action.

manual changeover An operation performed by a human operator to remove one set of equipment or *lines* from operation and to make another set available.

many-to-many call A *call* in which there is more than one *terminal* connected at each end of the *circuit*, thus allowing several *users* to speak and listen as though in conference. Contrast with *one-to-one call*.

mark In *telegraphy*, one of two possible line conditions occurring in a *signal*; the other condition being known as *space*. All *characters* in *alphabetic telegraphy* are made up of combinations of marks and spaces; i.e. patterns of *bits* representing the *binary digits* 0 and 1.
In a transmission sequence, the first bit of each character is always a space which acts as a *start pulse*. The final bit, or *stop pulse*, is always a mark and the line condition remains as a mark until another character is transmitted.

mark inversion ⇨ *alternate mark inversion*.

marker A *control* device which determines the switching path required in an *exchange* for a particular *call* and which operates the appropriate *crosspoints*.

master clock A timing device which generates *signals* to control events and has control over other *clocks* in the same equipment or the same *network*.

maximum frequency error The specified or recommended span of *frequency error* allowed in a specific system in order to remain within operating tolerances. For example, in *multiplexed channels* for *telephony*, a maximum frequency error of ± 2 Hz in a *carrier* system is a *CCITT* recommendation.

maximum justification rate The maximum rate at which *bits* can be inserted (or withdrawn) from a *digital signal* to make the rate of the signal conform to another desired rate. ⇨ *justification*.

maximum stuffing rate Same as *maximum justification rate*.

mean busy hour An uninterrupted period of 60 minutes in the daily operation of a system for which the total *traffic* is greater than for any other 60-minute period.

mean down time An expression used in assessing the reliability of a system, or system component, and referring to the average time for which a failure persists before it is corrected and the unit is operational once more. ⇨ *fault rate*.

mean holding time The average time for which *calls* occupy equipment in a given period. If, in a given period T, the system carries n calls and the durations of the calls are $h_1, h_2, h_3, \ldots h_n$ seconds, then the utilization of the system is:

$$\sum_{i=1}^{i=n} h_i$$

The mean holding time h is given as:

$$h = \frac{\sum\limits_{i=1}^{i=n} h_i}{n} \text{ seconds}$$

⇨ *traffic volume*.

mechanical interface The physical construction of the linkage between two devices; e.g. the pin connections and sockets connecting two units in order to carry *electrical signals* to control communication.

medium speed modem A device for connecting a *data terminal equipment* (*DTE*) to a communications line and capable of operating at speeds up to 2400 *bits* per second. ⇨ *modem*.

memory unit A storage device which stores *digital signals* until they are required for further processing; e.g. in a *terminal*, the memory unit is used to store *data* passing to or from the communications *line*.

mesh network A *network* in which *nodes* are connected to several other nodes; possibly every node is connected to every other. In this way there are a variety of paths for the transmission of *messages*. Contrast with *ring network* and ⇨ *network topologies*.

mesochronous signals Two *signals* whose significant timing intervals coincide.

message ⇨ *message signal*.

message feedback system A system in which the accuracy of *data transmission* is checked by automatically feeding received *signals* back to the *transmitting terminal* for comparison with the original *message*. The transmitting terminal automatically retransmits *erroneous blocks*. Same as *information feedback system*.

message information That part of any transmitted signal which represents infor-

mation to be conveyed from one end *user* to another.

message queues In a *store and forward system* the *transmission links* may be temporarily overloaded by *traffic* passing through them. *Messages* within the *network* are then stored in *computer*-controlled *exchanges*. Message queues are formed and messages released according to priorities when a suitable *routing* is available.

message signal That part of a transmitted *signal* which contains *information* intended for a *user* at a remote location. It is sometimes used to distinguish the message from *control information* required for the management of the message through the *network*.

message source The *terminal* from which a particular *message* is first transmitted; e.g. a particular *telephone* or *teleprinter*.

message switched exchange Message switched exchanges are used in *data networks*. In a message switched exchange it is not necessary for a point-to-point connection to be made between two *terminals*. In practice, the exchange has the capacity to store *messages* that it receives, until it is able to establish a path to the required destination. Unlike *circuit switched systems*, the *signalling* processes are not separated from the communication processes in a message switched system. In fact, the required destination of the message is stored as a *header* to the information contained in the message.

Messages are associated by the *destination address* with a particular exchange to which the *destination terminal* is connected. Thus, a message switched exchange has to examine the address, determine whether the destination terminal is connected directly to the exchange, and deliver the message to that terminal, or to another exchange as soon as the appropriate *channel* is available. Messages are transmitted one after another down each *link* (but ⇨

packet switching in which messages may be interleaved), regardless of whether the destination terminal is free. When any particular *node* (or exchange) in the network is overloaded, or when a particular terminal or link is unavailable, the messages are queued in store at one or more exchanges.

There is a *router* in each exchange which manages these queues, placing the messages into output *buffers* when appropriate links are available.

A router is usually realized as a *software* program operating at high speed in a general-purpose *computer*. The router has basically three functions: (a) collecting incoming messages, (b) determining message routing, and (c) sending messages to required destinations.

message switched system A system in which point-to-point connections need not be available to successfully transmit a *message* from sender to receiver. Instead, the *transmitting station* transmits the message to the *network* where it will be received and acknowledged by an *exchange* or *node* which will take responsibility for *routing* the message to the destination. ⇨ *message switched exchange*.

message switching The techniques concerned with transmitting *messages* over a *network* which includes *exchanges* (or *nodes*) which are able to store messages and forward them on when appropriate *circuits* are available. ⇨ *message switched exchange*, and contrast with *circuit switched exchange*.

message synchronization With synchronous *data transmission* systems, the beginning and end of each *message* is usually indicated by special *control* codes. These *codes* are identified by the *receiving station*, and it is thus possible to divide a stream of incoming *bits* into messages, *blocks* or *fields*. Examples of such codes are given under the heading *transmission control codes*.

message transfer The activity associated with making a *call* on a *switched circuit* is usually considered in three phases. These are: *call establishment, message transfer,* and *call clearing*.

Once a *transmission path* is established, the transfer of the *message signal* takes place. This phase is referred to as message transfer (or *data transfer*).

The concepts applied to these phases are the same for both *data transmission* and *voice telephony*, but the detailed methods of transmission differ.

MF signalling Abbreviation of *multifrequency tone signalling*.

microcomputer A class of small *digital computer* based upon the technology of *micro-electronics* which began to appear in commercial applications in the late 1970s. The first microcomputers were based upon 8-bit *microprocessor* central processing units, mounted on a board with *random access memory (RAM)* chips for program and *data* storage – the equivalent of *main memory* in a *main frame* or *minicomputer*. Also mounted on the board are *read-only memory (ROM)* chips which contain *software* providing programming and *operating system* environment. Usually microcomputers are programmed in an *assembler* language or a *high level language* such as *BASIC*.

In the simplest form, input is via a *keyboard*, and output is displayed on a TV monitor driven from a *UHF* TV *modulator* mounted on the board.

Typically, an early microcomputer has a minimum of 4K *bytes* of RAM, capable of expansion in 4K byte modules up to, say, 32K bytes, and from 4K bytes to 16K bytes of ROM.

Additional data storage facilities are available via an *interface* to a standard audio cassette player or a *floppy disc drive*. A small lightweight printer device is available.

Rapid improvements in *micro-electronic technology* will mean that microcomputers will soon have some of the features associated with earlier minicomputers.

In the early 1980s, the first 16-bit microprocessors are being introduced and a wider range of *programming languages* is available for microcomputers, as well as 200 megabyte *hard disc storage* systems.

First introduced as personal computers and education tools, microcomputers are now being used as *intelligent terminals* in business applications, as *front end processors* to interface minicomputers to *networks*, and to augment minicomputers by providing special arithmetic or *graphics processing* applications. ⇨ *main frame* and *minicomputer*.

micro-electronic technology The application of solid state technology in which semi-conductors, constructed of silicon, can be arranged to provide complex logical functions within very small components, designed, manufactured and tested by automated methods. The devices produced by this technology create opportunities for integration of many components and logic functions into exceedingly small physical size. The concepts of *large scale integration* (*LSI*) and *very large scale integration* (*VLSI*) have been introduced through this technology.

These developments continue to have a large impact upon the communications industry; they mean that systems can be constructed with high inherent reliability, with small physical space and electrical power requirements, and with economic benefits of automation in manufacture and testing.

microphone A *transducer* which converts *sound waves* into *analog electrical signals* in the *audio frequency band*.

microprocessor A small micro-electronic device providing a powerful central processing unit for a *computer* on a single chip with *interfaces* to enable it to be interfaced to various peripheral units for the input, storage and output of *data*. The central element of a *microcomputer*.

microwaves *Radio waves* having very short *wavelength* and occupying frequencies in the range above 1 GigaHertz (1 GHz = 1×10^9 Hertz). They are used to transmit *signals* along *waveguides* and for point-to-point directional radio links.

minicomputer This term applies to a class of *digital computer* which began to be introduced in the late 1960s and gained rapid penetration among *users* who wished to develop dedicated *applications* for multiple users, rather than establish large general-purpose *computers* to control all the applications in an organization. Minicomputers have been applied across a wide variety of commercial and scientific applications, and have been extensively used as *communications controllers* in *data transmission* and *message switching* applications. The precise distinction between minicomputers and *main frame* computers is difficult to define since advancing technology is increasing the power and scope of minicomputers. Typically, a mini would operate with a 16-bit central processing unit, have up to 32 communications *ports*, up to 64K bytes of *main memory* and up to 60 megabytes of *backing storage*. The models being introduced during the early 1980s include 32-bit central processing units, up to 200 communication ports, a minimum of 512K bytes of main memory, and a minimum of 250 megabytes of backing storage.

Minicomputers usually support a wide range of *programming languages* and are particularly suited to *distributed data processing* applications. ⇨ *main frame* and *microcomputer*.

minimum signal level Communications equipment is designed to respond to *electrical signals* of a particular strength. Often the power of a signal is *attenuated* over a *transmission channel*, and *repeaters* are used to boost the signal to the correct strength. The minimum signal level is the level below which a signal will not be acceptable to a particular class of receiving equipment.

modem This term is a contraction of *modulator/demodulator*. Modems are devices used to *interface* communications equipment (e.g. *terminals* and *nodes*) to a *transmission line*. There are a variety of modems and the evolution of the modem has seen a number of new technologies absorbed into the design.

Today, modems are generally classified into speeds:

low speed – up to 1200 *bits* per second
medium speed – up to 2400 bits per second
high speed – up to 9600 bits per second
  and beyond.

Most modems are designed to connect *data terminal equipment* (*DTE*) to *public networks*, but there is an increasing demand for communications networks covering large *in-plant* installations which do not use the public network. Modems for this purpose are known as *short-haul modems*, and they provide a means of interfacing *terminals* in *local mode*.

Early modems operated using *analog techniques* for transmission of data over *telephone networks*, but the increasing spread of digital technology allowing high-speed transmission with very low *error rates* has seen the development of modems which directly interface a *digital terminal* to a digital *transmission channel*. These devices are more strictly titled *network interface units*, but are called modems by popular convention. The term *data circuit terminating equipment* (*DCE*) is the term preferred by the *CCITT* to denote a device which interfaces a data terminal to a network.

modulated carrier wave A *signal* used to carry a *message signal* which has been imprinted upon it for the purpose of transmission. ⇨ *frequency division multiplexing*.

modulation *Low frequency* signals (e.g. *audio frequency*) cannot be effectively radiated through space. Thus, from the beginning of the history of *radio transmission* it became necessary to carry low frequency signals imprinted on to a *high*

frequency carrier wave. There are two principal methods for achieving this: (a) *amplitude modulation*, and (b) *frequency modulation*.

In amplitude modulation (*AM*), the frequency of the *carrier wave* is retained constant, but the amplitude of the carrier wave signal is varied in exact proportion to the *message signal* to be transmitted. The receiving equipment *demodulates* the signal by comparing it with a generated *carrier signal* of constant amplitude.

In frequency modulation (*FM*), the frequency of the carrier wave is not retained constant, but is allowed to vary about a reference point in proportion to the low frequency signal to be transmitted. The receiving equipment demodulates the signal by comparing it to a generated carrier signal of constant frequency.

Both forms of modulation are widely used. Frequency modulation is often used for *sound* radio broadcasts today, because it is far less likely to be disturbed by interference or *noise*. Modulation has become widespread in other fields of *telecommunications*. For example, in *telephony*, a single *transmission path* can be used to carry hundreds or thousands of separate *calls* using the technique known as *multiplexing*. A different carrier wave is modulated by each call, so that resulting signals are separated from one another by a frequency displacement which ensures that there is no mutual interference as they are transmitted along the same path.

Amplitude modulation is often used in *data communications* in preference to frequency modulation because it is more efficient in the use of *bandwidth*, but a special variant of frequency modulation is also used to transmit *digital information*. This is described under the heading *frequency shift keying* (*FSK*).

A third form of modulation is known as *phase modulation*. This is a system whereby binary values such as 0 and 1 are indicated by an alteration of the *phase* of an *alternating current* (*AC*) carrier signal. Thus, without changing amplitude or frequency, information is carried and the

105

message signal is detected by comparing the carrier with a *reference wave* to determine the angle of phase lag.

There are many variations of these basic modulation techniques; an example occurs in television *broadcasting*. The objective of such a mass system of communication is to have a cheap and simple *TV receiver* without appreciable *distortion* arising in the *demodulation* process. At the same time, it is necessary to economize in the bandwidth and power required in transmission. A form of amplitude modulation known as *vestigial sideband modulation* is used.

The normal process of amplitude modulation produces a *spectrum envelope* which contains two identical sidebands: one below the carrier wave and one above the carrier wave. The message signal is repeated in both sidebands. A great saving in power and bandwidth is possible by suppressing the carrier wave and one of the sidebands, using a technique known as *single sideband modulation (SSB)*.

This SSB technique is used extensively in communications, but it requires greater complexity at the *receiving station* to demodulate an SSB signal. In the case of television broadcasting, the *upper sideband* is transmitted along with part of the *lower sideband* known as the *vestigial sideband (VSB)* This simplifies the circuitry required in the TV receiver. ⇨ *multi-level amplitude modulation, variable amplitude modulation*, and *detection*.

modulation rate A term used to express the performance of a *circuit*, indicating the rate at which pulse changes occur. For example, if a *signal* consists of pulses of 20 milliseconds' duration, the modulation rate is $1 \div 0.02 = 50$ per second. This is also known as the *baud rate*, but does not necessarily refer to the rate at which information is transmitted. ⇨ *data signalling rate* and *data transfer rate*.

modulator A device which takes a *message signal* and converts it into another form suitable for transmission as a *radio signal*

or a signal to be combined with other message signals in a *multiplexed* transmission system. ⇨ general article on *modulation*.

monochrome display A television, or a *visual display* device, based on the principles of the *cathode ray tube*, which is able to display information in one basic colour such as black and white; i.e. not a full colour display.

monochrome receiver ⇨ *monochrome display*.

Morse code A *code* developed for the *electrical telegraph* by Samuel Morse, an early pioneer of electrical *telegraphy*. The code enabled *characters* to be represented as a series of electrical pulses of short and long duration which were generated by a manually operated key used to interrupt a *circuit*. These pulses were detected as sounds by a person wearing earphones at the *receiving station*.

Morse, Samuel An early pioneer of electrical *telegraphy* who, in 1838, succeeded in transmitting a *message* over 5 kilometres using a *telegraphic code* developed by himself. In 1844 this code, known as the *Morse code*, was used for the first public telegraph between Washington and Baltimore. Some years later, the Morse telegraph was recognized by the *International Telegraph Convention* as an international standard.

mosaic graphics set A set of *characters*, and the *codes* representing those characters, which are used in creating images for display in any *videotex system* which uses *alphamosaic coding*. The characters can be combined to form images of relatively *low resolution*, but, for example, cannot be used to draw fine lines and arcs.

most significant bit In any *binary number*, or unit of *binary coded information*, the highest order *binary digit*. In a communications system, this may not necessarily by the first *bit* transmitted.

MSB Abbreviation of *most significant bit*.

muldex Abbreviation used as a short-hand of *multiplexor/demultiplexor*. ⇨ *demux*.

multiaddress calling A system in which *data* can be *broadcast* to many *users* in a *network* by nominating individual *addresses*, or by use of a special *code* which identifies a predetermined group of addresses.

multi-block In *signalling system No. 6*, a group of 96 *signal units* transmitted as a *block* on the signalling *channel*.

multi-block synchronization unit A *signal unit* carrying information to synchronize a number of units of signal information transmitted as a *multi-block* on a signalling *channel*.

multi-cast address A *destination address* which relates to a specified group of logically related *stations* (e.g. a *closed user group*) or a set composed of all the stations on the *network*.

multidrop line A *communication channel* which services several *data terminals* at different geographical locations and in which a *computer* (*node*) controls utilization of the channel by *polling* techniques.

multidrop network A *network* in which the various *nodes* or *terminals* are positioned along a single *transmission link*, so that transmission must proceed between any two nodes via any intermediate nodes, with the controlling node positioned at one end of the *link*. ⇨ *network topologies*.

multi-exchange call A *call* which passes through more than one *exchange*; i.e. not an *own exchange call*.

multi-exchange connection Same as *multi-exchange call*.

multiframe A group of *frames* considered as an entity in a system, and in which the position of each frame is detected by reference to a multiframe alignment *signal*.

multifrequency tone signalling The traditional method of sending *signals* to a telephone *line* is by creating pulses by interruption of a *circuit*. A *telephone* dial or push-button panel can be used to cause this interruption which is known as *loop-disconnect signalling*.

Multifrequency tone signalling is a more modern method of *signalling*, in which each *digit* is represented by two tones of different frequency sent to the line. Each button of the telephone is connected to two *oscillators* which are activated when it is pressed. The oscillators are situated on the telephone and draw *DC* power from the *local line*. This method of signalling allows the *calling signals* to be generated more quickly than a telephone which uses loop-disconnect signalling. However, only certain types of exchange can respond to multifrequency tone signalling.

multi-level amplitude modulation A form of *amplitude modulation* in which the *amplitude* of a *carrier wave* is varied to represent digital information and in which certain predetermined levels of amplitude are used to represent a *binary number*; e.g. a separate level exists for 00, 01, 10 and 11. ⇨ *modulation*.

multi-level signalling Any method of transmitting *binary coded information* in which groups of *binary digits* (e.g. 00, 01, 10, 11) are represented by different signal status. For example, in *frequency shift keying* a number of frequency levels can be used to represent such conditions.

multi-link calls A *call* in which a *transmitting terminal* is able to transmit information to more than one *receiving station* at the same time.

multi-party connection A facility provided, particularly in *digital exchanges*, to allow three or more parties to participate in a single telephone *conversation*.

multi-phase modulation A method of *phase modulation* in which more than two variations of *phase* angles are used. Thus, more than two values can be represented. For example, in a four-phase system: $0° = 01$, $90° = 00$, $180° = 10$, $270° = 11$, In an eight-phase system, sufficient combinations are available to represent values for 3 *bits* per phase.

multiplex aggregate bit rate In a *time division multiplexing* system, the sum of the *bit rates* of the individual *input channels* plus the *overhead bits* required by the *multiplexing* process.

multiplexed Pertaining to any *circuit* or *transmission path* in which a number of *channels* are combined on to one physical path using the techniques of *frequency division multiplexing* (*FDM*) or *time division multiplexing* (*TDM*).

multiplexed channel A *channel* used to carry simultaneously a number of *message signals*, which have been combined by *multiplexing* them on to a *carrier signal*. For example, a *telephone* message signal occupies a *bandwidth* of 3 kHz, but several *telephone channels* can be obtained for simultaneous transmission over a single *broadband channel*. Thus, a channel having a bandwidth of 1 MHz would support up to 250 channels, each of 4 kHz using *frequency division multiplexing*. In such cases, telephone channels are each allowed 4 kHz in order to reduce the need for sharp cut-off *filters* when *demultiplexing*.

multiplexed traffic Relating to *message signals* which have been converted into a form, and combined in such a way so that hundreds or thousands of signals may be transmitted down the same *transmission path* at the same time. ⇨ *frequency division multiplexing* and *time division multiplexing*.

multiplexing A technique used to carry several hundreds, or thousands, of *message*

signals on a single *transmission channel*. Two main techniques are used. ⇨ *frequency division multiplexing* (*FDM*) – for *analog signals*. ⇨ *time division multiplexing* (*TDM*) – for *digital signals*.

multiplexor A device which enables a number of *message signals* to share the same physical *transmission channel* by using the techniques of *frequency division multiplexing* (*FDM*) or *time division multiplexing* (*TDM*) Such devices are used to improve the utilization of communication *networks*. A *transmission path* which is used in this way is known as a *multiplexed channel*.

multi-state signalling A system for handling *digital signals* in which several values can be encoded in a signal element. In most digital systems, information is transferred as a series of pulses representing the *binary digits* 0 or 1 according to the pulse polarity – all the pulses being of constant duration. In some systems, pulses may take on different values, according to the *amplitude* of each pulse, the duration of pulses remaining identical in each case. As an example, a system in which pairs of binary digits, 00, 01, 10 and 11 are represented by different amplitudes provides for four possible states for any pulse, and gives a higher *data signalling rate* than the more familiar *two-state signalling* method.

multistation DLC A *data link control* which performs error checking facilities on a number of *multiplexed* streams of *data*, interleaved upon the same *line* by *time division multiplexing* techniques. Used where a number of *nodes* use the same *leased line*, and the DLC includes link address fields and *control information* which assist the DLC to avoid conflict on the line.

music circuit A high-quality *circuit* having a *bandwidth* capable of handling a range of frequencies sufficient to carry *signals* over the whole *audio frequency band*.

mutually synchronized network A *network* in which all the *clocks* exert some degree of control on one another, and no single clock is dominant. Also known as *demo-* *cratic network*, and contrasted with *despotic network*.

MUX Abbreviation of *multiplexor*.

N

NAK ⇨ *negative acknowledge*.

narrow bandwidth Relating to a *channel* which can carry *signals* of only *low frequency* (e.g. *voice frequency signals*). ⇨ article on *bandwidth*.

national circuit A *circuit* which connects two *exchanges* in the same country.

national significant number An *address* which includes a *trunk* code and a *subscriber* number to provide a unique identification of a party.

National Television Systems Committee (NTSC) A committee appointed to develop the standards and principles of the television broadcast system used in the USA, and adopted also in Mexico, Canada and Japan. The system uses 525 picture lines and a 60 Hz *field frequency*. ⇨ *SECAM* and *PAL*.

nature-of-circuit indicator Information included in the *call establishment* process and transmitted on a *forward channel* to instruct an *exchange* of the type of *circuit* already used for the connection. This assists in the selection of an appropriate circuit for ongoing connection.

negative acknowledge (NAK) An international *transmission control code* returned by a *receiving terminal* to signify that a *frame* of *information* has been received but the frame is incorrect. Contrast with *acknowledge (ACK)*.

negative bit stuffing Same as *negative justification*.

negative justification The deletion of *digit pulses* from a *digital signal* to reduce the *bit rate* to a rate required. ⇨ *justification*.

network The basic objective of any network is to provide *access paths* between *users* at different geographical locations. A great variety of terminal types may exist in the network and a large number of *nodes* may be present. The network may have a number of possible *routings* for a *call*, depending upon the geographical distribution of the nodes (or *exchanges*), and the *circuits* linking these nodes. The network makes it possible for end users to be connected and to communicate intelligibly in spite of errors, difference in speed of operation, *protocol* and format.

network architecture A precise definition of the *protocol*, functions, and logical components of a *network* and how they should perform. Such architectures embrace the interaction of different levels of *control* which must be implemented at *transmitting* and *receiving stations* to maintain a coherent system of communication. An example is described under the entry *ISO reference model for open systems architecture*.

network attachment It is usual for authorities who operate *public telephone* or *data networks* to have control over the approval of *terminals* or other devices which may be connected to the network by *subscribers*. The object of this is to ensure that the electrical and functional characteristics of devices attached to the network do not impair the reliability of the network or impair the quality of service enjoyed by other subscribers.

network control A *network* may consist of a number of diverse *stations* operating to different *protocols* and with a wide geographic distribution. A most important element in any network is the set of functions which control the activation and deactivation of *access paths* between *trans-*

mitting and *receiving stations*. These functions include certain aspects concerned with the operation of the network including the management of recovery procedures in the event of failure of a *node* or *transmission link*.

Network control is sometimes centralized (i.e. resides in a single node) or decentralized (no single node dominant).

In a typical network, using *packet switching* technology, the following network control functions must exist:

1. Establishing the *transmission path* between nodes (e.g. providing onward dial-up information).

2. Maintaining *routing* tables to allow each node to select outgoing links for onward transmission.

3. Assigning authority for activation of *data link control* functions to appropriate stations.

4. Maintaining directories of all *users* in the system.

5. Setting up and deactivating the pattern of *dialogue management* requested by a user for a particular *call*.

6. Managing queues of requests and responses within a *session*.

7. Communicating network status to the operational staff and conducting tests and measurements of activity.

Network control is a vital consideration in any data network, but is particularly emphasized where combinations of protocols involving private and public facilities are to be utilized.

network control layer In a layered *network architecture*, this term defines that layer of logic which provides for *network control* between adjacent network *nodes*; e.g. the standard which allows for *packet* formatting and *virtual circuits* in an *X25* implementation. ⇨ the *ISO reference model for open systems architecture*.

network control phase During a *call* in a *data network*, any period during which a *data terminal* (*DTE*) exchanges information with the network to manage the set-up, progression or disconnection of a *call*.

network diagnostic controller A device which is connected to a *data communications* network to monitor and automatically test the quality of *lines*, *modems*, or other equipment forming part of a *network*. It will help to detect and isolate network problems, and, in the event of failure or poor performance in some aspect of the network, it will provide for the reconfiguration of the network from a central location without manual intervention at remote sites.

In some cases, monitoring can be carried out on secondary low-speed lines independent of the *primary channels* so that monitoring does not interfere with the main data paths.

Diagnostic information is usually presented on a *cathode ray tube* display, and often a *hard copy printer* is provided to give a permanent record of all tests.

Automatic alarm systems can be included in such systems to give visible and audible warning of major faults which have interrupted, or are likely to interrupt, key services.

network facilities In any *network* there may be a range of facilities available to a *user* making a *call request*. The selection of the facilities required for a *call* are defined when it is established. In many *data networks* this is defined as part of a *facility request*, which is a sequence of *characters* transmitted as part of a *message signal*.

network implementation The particular realization of a *network architecture* in *hardware* and *software*; i.e. the way in which the system builders have built a system to conform with the logical and *protocol* requirements of the architecture.

network interface unit A device used to connect a *digital terminal* to a *digital channel*, for sending or receiving *message signals* representing *data*. The input and output signals are in digital form, and the essential function is to balance the signals produced in the terminal with the standards expected on the *transmission channel*. ⇨ general article on *digital data transmission*.

network maintainability A quality desired in a *network* which allows for operation of the network under various conditions including scheduled and unscheduled maintenance. It also allows for planned extension of the network and its *traffic* capacity to be carried out without disruption of service to users.

network maintenance signals A special class of *signal* used to assist in the management and operational maintenance of a *network*. ⇨ general article entitled *signalling*.

network management Relates to the set of functions that perform control over the operation of a communications network; e.g. the selection and release of the various paths through the *network* to service a particular *call*, the management of all *recovery procedures* to ensure continuous operation of the network, and protection for *subscribers'* messages during failure.

The network management also includes maintaining a directory of all subscribers to the network and providing conversion from subscriber name to the *address* required. Another important function is the management of queues of requests and responses within the system and, not least, the interaction with the human *network manager*.

In large *public networks* there are also specific tasks performed in a *network management centre* including *traffic analysis*, *network performance*, *billing* and *accounting*, *configuration control* and *fault detection*. Also known as *network control*.

network management centre (NMC) An overall centre for control of a *network*, including: recording for *billing* and *accounting*; *traffic analysis* and *network performance*; *software* loading and maintenance of *nodes*; *configuration control*; and *fault detection* and diagnosis. ⇨ *network management*.

network management point A location in a *network* at which information regarding the loading, utilization and maintenance of the network is concentrated and analysed. Such a location may have human operators who are responsible for *network management* and who use the information presented to ensure the maximum utilization of the network under various operating conditions.

network management signals Special *signals* transferred to and from *network management points* to notify conditions related to the *circuits* and transmission and *switching equipment*. Concerned with *network management* rather than the progress of individual *calls*, and intended to improve utilization and *flow control*.

network manager A person who has responsibility for the maintenance of a *network* and for the level of service experienced by *subscribers*. ⇨ *network management*.

network operator An organization responsible for the operation of a *network*; e.g. a *common carrier* providing a service to business *users* by providing *circuits* and *switching equipment* to carry *data* over a *public data network*.

network performance The performance of a *network* is measured in a number of ways. First, a network should allow a *subscriber* to gain access to the network with a high probability of obtaining connection to the *local exchange* or *node*, and with a high probability of obtaining an *access path* to a particular *called location*. This aspect of service is referred to as *availability of service*. The quality and reliability of the service are also important considerations, and affect both availability and the delay across a network. Poor-quality *circuits* will result in noticeable delays or poor reception in *speech circuits*. Studies of *traffic* patterns are necessary to maintain a network capable of handling a *peak load* of *message signals.* Performance is one of the aspects of *network management.*

network terminating unit A device which *interfaces* one or more *terminals* to a *network* and converts *signals* to or from the form expected by the network.

network topologies The purpose of any *network* is to provide *access paths* for end *users* to communicate. In networks which involve computer *databases,* the purpose of such communication may be to obtain answers to enquiries addressed to the database, to update *records* in the database, to send *messages* between one user *terminal* and another, or to transfer sections of the database from one location to another.

The physical arrangement of the *nodes* and the interconnecting communication *links* in the network is dependent upon the requirements of the *application* and the geographical distribution of the users. Certain distinctive patterns can be discussed; these patterns are referred to as network topologies. Five such topologies are referred to below, and are illustrated in Appendix 11.

1. *Star Network*: this type of network is characterized by having a central node at which a controlling *computer* handles all functions involved in communication. The users all communicate to this centre directly along individual communication links. There may be intelligent concentrators at certain locations, but in principle all intercommunication between users takes place to and from, or via, the centre.

2. *Loop (or Ring) Network*: in such a network there is no central controlling computer, but instead an agreed *protocol* by which each node in the network communicates with others. One node may have certain overall administrative functions (such as collection and generation of performance statistics), but in principle each node directly communicates with adjacent nodes only, and transactions may be routed around the ring in any direction.

3. *Multidrop Network*: this is one of the simplest networks in which nodes are posi-

tioned along a single communication link. Such patterns are common where the cost of a *leased line* is a predominant expense in the construction of the network, and it is desired to utilize the link fully, rather than invest in additional lines. The controlling node occurs at one end of the link.

4. *Tree Network*: a tree network involves a hierarchy of *control* which can be represented as a pyramid of nodes, in which the controlling node appears to be the apex of the pyramid. The lower levels of the pyramid act as *concentrators* or distribution nodes for all communication to and from the controlling node.

5. *Mesh Network*: a network in which there are many interconnections between nodes, and many choices of possible *routing* for a message or transaction. Taken to extremes, every node would be connected to every other, but this tends to be expensive in terms of communication links, unless of course a dial-up service is available to connect nodes over a public network. ●

In practice, most networks are formed of some combinations of the topologies described above and, in fact, examination of the exact topology of the network is not necessarily the best way to understand its nature. It is more important to examine what the network has to do to meet the functions and volumes required by users and the method of control. However, the topology does suggest where problems of control may be found and is a useful concept in understanding the operation and requirements of *networking* systems.

network user address (NUA) An identity number given to a *subscriber* to a *network* service to identify the *address* of a location from which the subscriber operates, or the address to which bills and utilization information are provided in respect of a given group of *users*; the exact use is dependent upon the service concerned. ⇨ *network user identity* and *sub-address.*

network user identity (NUI) An identity number given to a *subscriber* to a *network*

113

service once he has been registered as an authorized *user*. Very often an N U I relates to a specific terminal *address*, but in retrieval systems it can relate to a specific person who may gain access to the service from several *terminal* locations. On gaining entry to the service, a user may have to provide the N U I in order that the network can validate the use to be made of the service. It is also common for N U Is to be allocated under the umbrella of a *network user address* (*N U A*) which is the registered point at which a subscriber will accept bills for utilization of the service by associated N U Is.

networking The techniques and principles concerned with maintaining a system of communication in which *users* in different locations are able to communicate with one another (for the duration required for a particular *call*) by establishing *transmission paths* through a series of *exchanges* (or *nodes*). ⇨ *network architecture*, *network control* and *network implementation*.

NMC ⇨ *network management centre*.

no charge answer signal ⇨ *answer signal, no charge*.

node A point in a communications network in which a number of *communication channels* may converge and wherein some switching or *control* functions take place. The term is commonly used in *data networks*.

Nodes are generally computer controlled and may have a variety of functions according to the *architecture* and purpose of the *network* in which they exist. For example, a *telephone exchange* is a node; a *packet switching exchange* is also a node. Some nodes may be simple *concentrators* or *multiplexors*, whereas other nodes will have important *network control* functions. They may contain structures for receiving *messages*, controlling *routing*, and scheduling the transmission of messages via the *transmission links* to other nodes. They

may also store messages and control the maintenance of *message queues*.

node-to-network protocol A *protocol* which regulates the *interface* of a *node* to the next adjacent node in the *network*, to be contrasted with an *end-to-end protocol* which regulates transmission from one end user to another. For example, in the *X25* architectures, an end-to-end protocol exists to set-up *virtual circuits*, but thereafter the *addressing* is between each end user node and the network.

noise Any interference to a *signal* picked up along a cable and caused by the operation of electrical equipment; or, for example, in *radio transmission*, by a competing transmitter operating at, or near, the same point in the *frequency spectrum*.

Other examples include noise from the *amplification* of signals arising from electron agitation in *conductors* or resistors. This phenomenon varies with temperature. This is known also as circuit noise, thermal agitation, *white noise*, *gaussion noise* and *random noise*.

Crosstalk and *quantization* noise are also forms of noise arising in communication *links*. ⇨ *signal-to-noise ratio*.

nominal overall loss *Electrical signals* transmitted over *circuits* are attenuated (or weakened) by resistance in the *transmission medium*. This is known as *loss*. Losses can be attributed to the various components which make up the overall *transmission path*, and the losses are cumulative. Thus, an overall loss, or *insertion loss* as it is also known, can be determined. The loss varies with frequency, and it is usual to use a test frequency of 800 Hz in specifying and measuring losses. The nominal overall loss is the loss expected in a circuit of defined quality and length. Also known as *nominal total attenuation*.

nominal total attenuation In all communication *networks* there is a tendency for *message signals* to become impaired by loss of signal strength over distance – this is known as *attenuation*. In planning the implementation of a network or *transmission system*, a nominal value for attenuation is considered. For a given *call*, this nominal total attenuation is the accumulation of the nominal values arising in each *trunk* or *local circuit* used for the call.

Attenuation can be dealt with by use of *amplifiers* and *four-wire circuits*, and this must be done to ensure that individual trunks do not introduce an unwarranted level of attenuation. These matters are considered when planning transmission systems. Also known as *nominal overall loss*.

non-associated signalling Relating to a system in which *signalling* information is transmitted on one or more *channels* specifically designated to handle *signals* for a group of *circuits*. Contrasted with *channel associated signalling* wherein signals are carried on (or associated with) each individual *transmission path*.

non-directional antenna A transmitting *antenna* used to *broadcast* information more or less evenly in any direction, or a receiving antenna sensitive to *radio waves* arising from any direction.

non-director exchange ⇨ the description of *director exchange*.

non-linearity An effect experienced in processing *electromagnetic waves* in which various components of an input *signal* may be treated differently, causing *distortion* of output. For example, the *gain* or *loss* produced in a system may vary significantly across different frequencies in the *audio frequency band*. If steps are not taken to compensate, serious audible distortion may arise on output to the system *users*. These problems arise in *radio frequencies* also, and have to be considered in the designs for all kinds of transmission and receiving equipment. ⇨ *distortion*.

non-linearity distortion A form of distortion which arises in processing *electromagnetic waves* in which the *amplitude* of output produced is not proportional to the input across the *band* of frequencies being processed. ⇨ *non-linearity*.

non-revertive error checking A system of *error checking* in which a separate *channel* is used to report on errors detected at a *receiving terminal*; the main channel being used continually for *message* transmission. Contrast with *revertive error checking*.

non-spectral colours Colours produced by combination of *electromagnetic waves* from sources which simultaneously radiate energy from different *bands* of the visible light spectrum. Contrast with *spectral colours*.

non-synchronous network A *network* in which the *clocks* are not synchronized.

non-uniform quantization A *quantization* method in which the intervals of quantization are not equal.

NTSC Abbreviation of *National Television Systems Committee* of the USA, which developed the standards and principles of the television broadcast system used primarily in the USA, Mexico, Canada and Japan. The system uses 525 picture lines and a 60 Hz *field frequency*. ⇨ *SECAM* and *PAL*.

NUA ⇨ *network user address*.

NUI ⇨ *network user identity*.

NUL A special *character* used in a *data communication code* to represent a blank medium; e.g. for filling space in a *paper tape* when there is no information to be recorded.

number A *code* used to denote the *address* of a *subscriber*; e.g. a *local number* in a *telephone network* or an *international number*.

No. 6 exchange An *exchange* equipped with *stored program control* and able to operate in accordance with the specification of *signalling system No. 6*.

nyquist flank A region forming part of the *spectrum envelope* for a transmitted television *signal* using *vestigial sideband modulation*. This region is shaped on either side of the *vision carrier* and has the purpose of keeping *distortion* within acceptable limits.

octet A unit of data consisting of 8-*bits* (*binary digits*). For example, *packets* in a *packet switching system* are usually transmitted in multiples of 8-bits; e.g. the packet size may be expressed as 1024 octets, meaning 8,192 bits.

octet timing signal A *signal* in a *packet network* which is used to identify the first *bit* of an 8-bit sequence (or *octet*) occurring in a continuous sequence of octets.

odd parity check A check performed upon a unit of *data* where parity *bits* have been added to the data for *error control* purposes, and where the error control scheme is constructed to create odd parity in the rows or columns of bits which make up each *block* or *character*. ⇨ *even parity check*.

off-resistance The electrical resistance presented by any switch or relay in the off position.

offered traffic The volume of *traffic* which would arise if all *calls* required by *users* would be accepted by the system. Contrast with *carried traffic*.

office A term used in the USA to denote an *exchange*; i.e. a centre in which *switching equipment* is installed in a *public network*.

oligarchic network A *network* in which most of the *clocks* which establish timing for events are under the control of a few of the clocks. It is an example of a *synchronized network*.

on-line processing A term used in *data processing* to denote any system in which the end *users* are directly connected by *circuits* to a *computer* running an *application*. The users are able to interact with the

system at any time when they wish to process a transaction. Contrast with *batch processing*.

on-line program development A method used for the development of computer programs in which programmers are on-line to a *computer* which contains *compilers*, editors and testing aids, and which enables the programmer to write, check, test, correct and update programs.

on-line system A *data processing system* in which all functions are performed under control of a computer processor, including the preparation and entry of input *data*, and the transmission and distribution of output results. Transactions are often dealt with as they arise in the normal course of business with minimal delay, rather than being batched for processing at a convenient time. ⇨ *real-time system*.

one-to-many calls A *call* in which a single *terminal* is connected by a *circuit* to a number of user terminals at distant locations. Thus, it is possible to have several *users* involved in a *conversation*. Contrast with *one-to-one call*.

one-to-one call A *call* in which two *terminals* are connected for *message* transmission, but only one *user* terminal is connected at each end of the *transmission channel*. Contrast with *one-to-many call* or *many-to-many call*.

one-unit message A complete *message* which is transmitted within a single unit of *data*; e.g. a *control* signal.

open systems architecture A design for a *data network* which fulfils the objective of allowing any *user* to communicate with any other user, regardless of the types of *com-*

puter systems and *terminals* in use. Such a system is a theoretical concept which is being sought by international standards organizations, and which will require the co-operation of equipment vendors, communications authorities and corporate users.

Such a system would require standard conventions to be used throughout the world, including the communications aspects and the processing aspects which take place in computer *applications*. An open system concept does not imply that a particular *computer* will automatically be obliged to receive *calls* from other such systems; it refers to the potential for such systems to communicate with one another if they agree so to do.

The article headed *ISO reference model for open systems architecture* explains in more detail the nature and structure of such systems.

open systems interface A specification of a standard method of *network* interconnection which allows any *user* to communicate with any other, regardless of the origin of the equipment and *software* involved. ⇨ *open systems architecture* and *ISO reference model for open systems architecture*.

operating system A special program permanently resident in a *computer* or a *communications controller* to control the *hardware* and *software* resources and to supervise the running of other programs, including *user applications*.

optical fibres Optical fibres provide a *bounded transmission medium* along which information is carried as pulses of light. The main application of this technology is to transmit *digital signals* for *data* or telegraphic communication. The medium provides a *broadband transmission channel*, and the signals can be carried over significant distances without *distortion*. In this respect, optical fibre channels have advantages over *co-axial cables* which hitherto

have provided the principal means of transmission in *telecommunications*.

An optical fibre is usually about one tenth of a millimetre in diameter and consists of a central core of glass (silica), surrounded by a sheath of similar material, but with a lower refractive index than the central core. Light energy transmitted along the fibre is gradually lost but, with fibres in use today, 10 per cent of the energy transmitted can be delivered over a distance of 3 to 4 kilometres, and there are expectations that this distance can be extended to 50 kilometres, without the need for *repeaters* to regenerate the signal.

Information to be transmitted must first be generated as electrical *digital pulses* representing a series of ones or zeros. This signal is used to activate a laser light source, which generates the pulses of light which travel along the optical fibre. The light is detected at the far end and converted back to electrical forms.

The great advantage of today's optical fibres is that they provide an almost distortionless medium over 10 kilometres with *data rates* of up to 8 million *bits* per second on a single fibre.

A great deal of development is being undertaken on the composition of fibres, which presently allow transmission of light in the infra-red region. It seems likely that very large *bandwidth* systems will be operated, in future, for applications in both domestic and business communications – the impact would be truly revolutionary.

Most offices and private homes are today connected to national communications networks by a simple wire transmission channel affording a bandwidth of only 3 to 4 kHz. Optical fibres provide an enormous increase in the bandwidth and would allow a great range of services, involving the transmission of wideband *video* and digital signals directly into every home and office.

In the immediate future, optical fibres will be used in the replacement of co-axial cable routes between exchanges in city areas, and then be extended on inter-city routes. Co-axial systems require repeaters every 2 to 3 kilometres, and the introduction of optical fibre systems will bring very significant savings in the size and cost of equipment and improve the utilization of underground ducts and cableways.

Optical fibres represent one of the most significant technological improvements in the field of telecommunications.

optical waveguides A *waveguide* which carries *message signals* as light signals along a *bounded medium* constructed of glass fibres. Present developments using *infra-red light* indicate that vast *bandwidths* are available; e.g. up to 100 THz. ⇨ general article on *optical fibres*.

originating traffic This term defines the use made of a particular *terminal* to make *outgoing calls*, and contrasts with *incoming calls*, referred to as *terminating traffic*. ⇨ *traffic*, *instantaneous traffic*, *traffic volume* and *calling rate*.

oscillator A device containing electrical *circuits* which can create *alternating current* signals at a pre-specified frequency or at a selected frequency. Used, for example, to generate *carrier waves* in a *transmission system*.

out-band signalling A system of *signalling* in which *signals* are conveyed outside the *band* of frequencies normally used for *message* transmission, but are carried on (or associated with) the *transmission channel*.

out-of-band components Elements which arise as stray signal currents arising from *message signals* but transferred on to *transmission channels* outside of the *frequency band* normally used for message transmission. Although they may not interfere directly with messages, they may be designated as harmful if they have any interference with other frequencies used in the monitoring or maintenance of transmission facilities.

out-of-order signal A *signal* sent on a *backward channel* to indicate that the *terminal* of a *called subscriber* is faulty.

out-slot signalling *Signalling* information associated with a *digital channel* but transmitted in *time slots* not within the slots used for message *bits* within the channel time.

outgoing call Any *call* which is routed to a distant *address*; e.g. from an originating *data terminal* (*DTE*) or from a *switched telephone exchange*.

outgoing lines The *circuits* which are used to output *message signals* from a *switching equipment* to a distant *address*.

outgoing trunks The *trunk lines* connected to an *exchange* for the purpose of carrying outgoing *traffic* to another exchange.

overall loss The sum of all the *losses* that arise end-to-end in a *circuit*, and usually measured in *Bels* or *decibels*.

Also known as *insertion loss*.

overflow A condition which occurs when a *call* cannot be connected via a particular group of *circuits*, and an automatic *routing* is arranged through another group in which a free circuit exists.

overflow bid An attempt to *route* a *call* through a particular *circuit group* in which the attempt has to be satisfied by a route through an available circuit on another circuit group, the first circuit group being fully loaded and unable to accept the bid.

overhead bits Signal elements which are added to *information bits* as part of the management of a communications process, and not intended for transmission to the end *user*.

overlap
1. An erroneous condition in *facsimile transmission* in which the scanning process causes elements of the picture to be overlapped.

2. An overlap operation occurs in certain *signalling systems* in which *address information* is transmitted before a *calling subscriber* has finished dialling the address.

own exchange call A *call* between two *subscribers* connected to the same *local exchange*. Contrast with *multi-exchange call*.

P

packet A relatively small unit of *data* (up to, say, 8,000 *bits*), transmitted over a *packet switching network* as part of a *message* to be transferred from one *user* to another. Each packet includes information as a *header* which identifies the *destination address* and the sequence of the packet within the overall message. Packets travel independently of one another, and perhaps by different routes, but are reassembled as a coherent *message* at the *receiving station*.

There are different types of packet; some carry user information (e.g. data packets), but others (e.g. *call request packets*) are used to request the *network control* to perform operations to establish or clear *calls*. ⇨ *packet switching network*.

packet assembly The process involved in converting a *message signal* into *packets*. ⇨ *packet switching network*.

packet delay A measure of the performance of a *packet switching network* is the time taken to transfer *packets* from the transmitting to the receiving location. If the network is congested, the packets are stored temporarily until a *virtual circuit* is available. In interactive systems, the delay may be noticeable to the *users* and affect the quality of service. Generally speaking, the average delay should be of the order of 200 milliseconds for interactive data users and 600 milliseconds for bulk *data transfer*.

packet disassembly The process involved in reassembling a *message signal* from *packets* received over a *packet switching network*.

packet level protocols Formal rules which define the structure of commands, responses and *bit patterns* concerned with the handling of *packets* of *information* in a *packet switching network*.

packet mode terminal A *data terminal* (*DTE*) which has the necessary logical structures built into the *hardware* and *software* to enable the terminal to format and transmit/receive *packets* to and from a *packet switching network*.

packet network ⇨ *packet switching network*.

packet sequence In *packet switching systems* each *message signal* is broken into a number of smaller elements for the convenience of transmission across the *network*. These elements, known as *packets*, may not be transmitted in strict sequence and each packet contains identifying and sequence information to enable the message to be reconstructed.

packet switching exchange (PSE) A *node* in a *packet switching network* capable of carrying out the full range of packet switching operations, including managing *transmission lines*, switching *packets* and performing *packet assembly* and *packet disassembly* functions, as well as all command functions to set up and clear *calls*.

In such a network there may be other nodes which simply provide access to the network for remote *subscribers* by concentrating *users* on to a transmission line having access to a PSE. These are known as remote *access points* (*RAP*).

packet switching network In a *circuit switched system* a physical *circuit* is made between two *terminals* for the duration of a *call*. With high-speed circuits carrying *digital data*, a much greater resource utilization can be achieved by sharing paths through the *network*. Packet switching systems proved a method to achieve such improvements in use of transmission resources.

To allow such path sharing, data is transferred through the network in packets which include, apart from the data itself, *addressing* and sequence information to control the progress of the packet in the network. Packets are relatively small units of data, consisting of, say, 8,000 *bits*, and several packets may be needed to complete a message.

If at any time the arrival rate of packets exceeds the capacity of any *node* in the network, the packets are queued in *computer* stores and forwarded when transmission time is available.

Packet networks provide a *virtual circuit*, which appears to the *users* as a permanent connection between two terminals but which, in fact, is shared with other users.

The logical structure of the packets is determined by the *protocol* of the system and has no relevance to the structure of the message to be transmitted. The *transmitting* and *receiving stations* have to be designed to *interface* with this protocol and may be responsible for *packetizing* and *depacketizing* the *user data*. This operation may also be performed in *packet switching exchanges* to which *subscribers* are connected.

A packet switching system is an example of a *store-and-forward system* and, as such, it can be adapted to provide speed, *code* or *protocol conversions*, enabling a number of dissimilar terminals to operate within the same network.

Originally, packet switching was designed to provide a more efficient method of transferring data over networks, but it is also possible to transfer digitized *voice* provided the end-to-end *packet delay* is not more than 200 milliseconds. ⇨ *voice packet switching*.

packet switching service A service provided by a service operator to make available the facilities of a *packet switching network* to registered *subscribers*.

packet switching system ⇨ *packet switching network*.

packetizing The process by which a *message* is broken up into smaller segments or *packets* for transmission to the *line* in a *packet switching network*. Each packet contains *addressing* information to identify the *user* to whom it is intended, the transmitting user, and the packet sequence.

pad A *software* routine, or a system element, which deals with *packet assembly* and *disassembly* in a *packet switching system*; i.e. it assembles the *user data* into *packets* complete with *header* and identifying information used to control the correct *routing* and sequencing of packets during the transmission process.

page A unit of information to be printed or displayed for a *user* of an information service. For example, a page retrieved using a *videotex system*. Also referred to as a *frame* but, in the British system known as *Prestel*, each page can consist of up to 26 frames, each of 960 *characters*.

page charge The *tariff* charged to a *user* of a *videotex system* by an *information provider* for access to a *page* of *information*.

page store A *memory unit* capable of storing a display of *information* for a *page* of *text/graphics*. For example, the memory unit in a *videotex terminal*. Also known as *frame store*.

paired-disparity code A *code* in which *bits* or *characters* may be represented under a given set of rules by two different conditions. An example is given under *alternate mark inversion*.

Also known as an *alternating code*.

PAL Abbreviation of *phase alternation by line*. A colour television broadcasting system developed first in West Germany and the United Kingdom. PAL uses 625 picture lines and a 50 Hz *field frequency*. ⇨ *SECAM* and *NTSC*.

PAM Abbreviation of *pulse amplitude modulation*.

paper tape ⇨ *punched paper tape.*

paper tape punch A device for recording *data* as holes in *punched paper tape*. It may be designed for off-line preparation of *messages*, or to record messages received over a *communication channel*.

parallel attribute coding A system for coding in which the attributes of a *character* to be displayed (e.g. its colour, height, or flashing/steady indicator) are stored as *codes* but not serially in the *page store*. Thus, the attribute characters do not result in a compulsory space in the *text* as is the case with *serial attribute coding*.

The parallel attribute system is more flexible and allows more information per *page*, but the terminal logic is more complex than that required for serial coding, and serial attribute coding can cope with most practical situations.

parallel data transmission A method of transmission in which the *bits* which make up a *character* or *byte* pass simultaneously along separate paths, rather than in *bit serial* mode.

parallel mode Pertaining to the transfer or transmission of all the *signal* elements of a *character* or *byte* at the same time along independent paths. Compare with *serial mode*.

parallel-to-serial converter A device which receives *characters* (or some similar unit of *data*) as a set of simultaneous signal elements along separate paths, and then presents these elements in serial form to an outgoing path. It may automatically generate *start bits* and *stop bits* for each character.

parallel transmission All forms of *data transmission* are serial in nature; i.e. *bits*, *characters* or *frames* are sent one after another along the *transmission path*. However, over very short distances, or within an electronic device, it is sometimes arranged that a separate signal path is available for each constituent bit of a character. This is known as *bit parallel transmission*.

parity check A check made upon the rows or columns of *bits* forming a *character*, *block* or *frame* of *data*. This is made as part of an *error control* procedure in which parity bits are appended to data prior to transmission deliberately to create an *even* (or, in some systems, *odd*) parity.

partial transmission A technique for entering *data* over a *network* in which an operator enters variable data on to a pre-determined form format displayed on a *visual display* unit. When the operator transmits the form, only the variable data is sent to the *line*. Also known as *split screen transmission*.

partitioned database In a system in which a geographically distributed population of *users* requires access to update or retrieve information from computer *files*, there are certain choices in deciding where to locate the *database*. These choices include: (a) a single central location in which all updating and enquiry takes place; (b) a central update centre, but with remote replicated versions of the database for local enquiry; or (c) a database in which information relevant to certain regions is updated, maintained and stored in each region.

The latter is referred to as a partitioned database and would be chosen where there is a need for an accurate and up-to-date version of the files in each region, and where an economic analysis of *traffic* flows within the total system, and within regions, supports the argument for such partitioning across network *nodes*. ⇨ *shared database*.

passive broadcast medium A physical medium, such as a *co-axial cable loop*, which allows *frames* of *information* to be propagated for reception by any of the *stations* connected to it. No attempt is made to *route* frames to specific *addresses*,

123

but each station includes a *frame grabber* which looks for frames addressed to it.

path The complete set of resources which provides a connection between two points for the purpose of transmitting and receiving a *message signal*. Paths may be linked by switches (*exchanges*) to form a *switched connection*. Where the method of transmission includes a separate *channel* for *go* and *return* directions, these are considered to form a single path.

path control An element in a *node* of a *packet switching system*, which is responsible for *routing* packets to the correct *data link control* (*DLC*) for transmission to another node, or to the correct *station* attached to the node itself. It is also responsible for *packetizing* or *de-packetizing*.

path independent protocol (PIP) A method used to transmit *information* over a *packet switching network* in which no specific path is established and each *packet* may, therefore, be *routed* independently of other packets in the same *conversation*. Contrast with *fixed path protocol*.

pattern of intermittency A term used to describe the nature of transmission between two *users* in a *data network*. For example, a pattern which entails transmission of a single information *packet* from one user to the other is known as a *datagram*. A *dialogue* which entails transmission in both directions to complete a series of connected functions is known as a *transaction*, and a dialogue in which a number of transactions are progressed is known as a *session*.

PBT Abbreviation of *push button telephone*.

PCM Abbreviation of *pulse code modulation*.

PDI Abbreviation of *picture description instructions*.

PDN Abbreviation of *public data network*.

peak load The maximum *traffic* experienced by a *network*, or planned for in the operation of a network. In *public networks*, the peak load occurs at the *busy hour*. In *private networks*, such as in *data processing* applications, the peak load may occur after a system failure, when the traffic which has been suspended or delayed is restarted.

peak volume Relating to *traffic* studies in communications *networks* and defining the number of simultaneous *calls* being conducted at the busiest period in the operation of a system.

peer interaction A concept which derives from considering the related functions performed at each end of a *data transmission* channel, in which components of the system act in pairs. For example, the *modems* at each end communicate with one another, regardless of information about the *text* or information for controlling the *link*. The *DLC* elements at each end manage the transmission and *error checking* without regard to *modulation* and *demodulation* performed by the modems. The *application programs* are concerned only with the text transmitted.

There is an implication that in a *network* with a totally peer style of interaction, no one *computer* has total responsibility for *network management*.

penetration of service A measure of the degree to which a particular service is accepted by the *target user population*. It is measured by the number of *users* from the target population who have installed a *terminal*. For example, the penetration of the *telephone* service into the residential market can be expressed as the average number of telephones per home.

per-call unit A class of *control* device which is associated with a particular *call* while that call is in progress; e.g. a device to monitor a *call answered signal*.

per-line unit A class of *control* device which is permanently associated with a particular *line* and cannot be shared; e.g. a device attached to a line to recognize *call request signals*.

per set-up unit A type of *control* device associated with a *call* during the period of *call establishment*.

percentage overflow Derived from the ratio of *calls* which are required to *overflow* (known as *overflow bids*) to the total number of calls bid to a particular *circuit group*.

perception data Information collected directly from the human *users* of a communication system as a result of tests conducted to find factors which make the service attractive to users, and which imparts satisfaction to them.

peripheral attachment Any device (such as a *data printer, floppy disc* or *paper tape punch*) which can be connected to a *data terminal* to provide additional methods of inputting, storing or outputting *data*.

permanent copy A *message* produced on a physical medium; e.g. a printed message. ⇨ *hard copy*. Contrast with *transient copy* or *soft copy*.

permanent virtual circuit A *virtual circuit* is one which appears to be a point-to-point connection between two *terminals* in a *packet switching network*. In fact, part of the path between the two terminals is shared by many other terminals using *time division multiplexing* techniques. A permanent virtual circuit occurs where terminals are permanently associated via a virtual circuit and, in concept, this is equivalent to a *leased circuit* providing permanent point-to-point connection between two terminals; i.e. no *call set-up* or *clearing* procedures are needed. ⇨ *packet switching* and *public data networks*.

person/machine interface The *protocol* which a human *user* must understand and operate in order to use a particular device for its defined purpose. In designing a system, it is an objective to make this protocol as easy and natural to the *user* as is possible.

phase Any *signal* based upon *alternating current* can be represented by a time graph which shows the variation in the voltage of the signal over time. Two signals of identical frequency and identical voltage *amplitude* can still differ from one another in that they may be out of phase; i.e. one signal voltage lags behind the other. In *phase modulation*, signals which are 180° out of phase are used to represent opposite conditions (e.g. the two conditions 0 and 1 in a system of *binary notation*). In theory, there are infinite variations of possible phase but, in practice, it is expensive to detect many variations economically. Phase modulation systems involving more than two phase angles are, however, used in *telecommunications* to represent different signal values. ⇨ *multi-phase modulation*.

phase alternation by line (PAL) A television broadcasting system developed in the United Kingdom and West Germany which uses 625 pictures lines and a 50 Hz *field frequency*. ⇨ *SECAM* and *NTSC*.

phase inversion A technique used in *phase modulation* in which the binary values 0 and 1 are represented by *unit signal elements* commencing with a change of signal phase in a specific direction. ⇨ *phase*.

phase modulation A method of *modulation* used to represent *binary information* in which the *phase* of the *carrier signal* is continually inverted to indicate the *binary digits* 0 and 1.

Two methods of phase modulation are: (a) *differential phase modulation*, and (b) *fixed reference phase modulation*.

In case (a), a change in phase simply indicates a change from 0 to 1 or 1 to 0. No *reference wave* is required to detect the signal.

In case (b), the process is as follows: a 0

digit is indicated by a group of contiguous cycles beginning from a *phase inversion* in one direction, and the 1 digit by a group of contiguous cycles beginning from an opposite phase inversion. The *demodulation* of the original signal involves the *detection* of different phase relationships.

phasing In *facsimile* systems, a process which takes place at the *receiving terminal* to ensure that a transmitted picture is correctly positioned on the recording medium. The process entails achieving coincidence between the significant points of the scanning field at each end of the *transmission line*.

phosphorescence A form of light radiation which continues after external excitation has ceased. The after-glow effect which is produced by the properties of phosphor materials after they have been excited by the action of an *electron beam*. For example, the light from a TV tube is created first by *fluorescence* which arises when the beam strikes phosphor materials on the tube surface; this light fades slowly due to the properties of the phosphor; e.g. it takes 4 milliseconds to fade to 5 per cent of its original value.

photograph facsimile telegraphy A form of *telegraphy* in which *signals* containing information relating to continuous tonal densities are transmitted to permit a faithful reproduction of any photographic image over a communications *circuit*. Compare with *document facsimile telegraphy* and *alphabetic telegraphy*.

photometric power The electromagnetic energy per second flowing from a light source measured to account for the human perception of such energy, which varies by sensitivity to different *wavelengths*. ⇨ *visibility function*.

photometry The science concerned with the human perception of light energy radiated from a source. This would include light emitted directly by sources, as well as light emitted as reflection, and incident light. Compare with *radiometry*.

photopic response The average response curve obtained by observing human visual perception of *electromagnetic waves*. ⇨ *visibility function*.

physical address A unique identification associated with a particular *station* on a *network*.

physical channel The physical medium (e.g. *co-axial cable*) used to convey *signals*, and having particular physical and electrical properties to meet the specified *physical link layer* of a defined *network architecture*.

physical control layer ⇨ *physical link layer*.

physical interface The specification of the *electrical signals* needed to establish, maintain and clear the connection between a *terminal* and a *data circuit terminating equipment* (*DCE*). Compare with *mechanical interface* and *logical interface*.

physical level interface In a layered *network architecture*, the lowest level of *control* responsible for actuating, maintaining and deactivating the *links* between a *terminal* and a *network* is called the *physical link layer*. The specification of the electrical and physical connections at this level is known as the physical level interface.

physical link layer In a *network architecture*, this level of *control* represents the lowest level in the *ISO reference model for open systems architecture*. This layer is dependent upon the physical medium for the *transmission channel*. It includes physical and electrical *interfaces* to the channel, including *data encoding*, timing and voltage levels. Compare with *data link layer*.

picel ⇨ *picture element*.

picture description instruction (PDI) A *code* which has been formed by compressing a *high resolution* picture definition to enable detailed and complex graphic images to be transmitted efficiently over a narrow band channel. The PDI is able to regenerate the original picture. An example of such a code is described in the entry for *alphageometric coding*.

picture element   In a television system, or in any *visual display* system used to display *text* and *graphics*, a picture element is the smallest discrete element at which a component of an image can be resolved. The number of horizontal lines determines the *vertical resolution* of the image, and in television systems 625 lines are recommended by *CCIR*. For various reasons, the vertical resolution is somewhat less than 625; in practice, for the UK system it is around 402 elements. The *horizontal resolution* is less restricted and is 572 elements.

Thus, in the UK television system, the resolution of a *monochrome display* is determined by a matrix of 402 by 572 picture elements, each of which can be considered as a black or white dot forming the overall picture.

In a visual display unit used for *data processing*, information to be displayed is created in a *memory unit* as a matrix of *bits* corresponding to the resolution possible by the properties of the screen. Individual *characters* to be displayed are composed of groups of picture elements, say 12 to a character. Fewer elements may be used, but this will reduce the quality of the character display. 3 or 4 lines may be used for each character row.

The arrangement of phosphor dots on the screen also imparts the resolution possible. The higher the resolution desired, the more dots are required.

The term *picel* (a contraction of *pic*ture *el*ement) is often used.

picture frequency   The frequency with which a complete set of *lines* representing a picture is displayed in a television system. Measured in cycles per second (Hz), the picture frequency is normally exactly half the *field frequency*. Typical picture frequencies are USA: 30 Hz; *CCIR* 625-line

system: 25 Hz. ⇨ *interlaced scanning* and *video signal*.

picture phone   A telephone device which presents a picture of the calling parties as well as the usual sound transmission. An example is a broadband teleconferencing system in which full motion images can be transmitted.

Such devices require expensive terminals and broadband transmission links, and the consequent cost of these facilities has probably impeded the growth of these systems versus voice telephone systems.

Picture Prestel   The name given by *British Telecom* to a *videotex system* first exhibited in March 1980, and demonstrating the storage and display of photographic images as part of a videotex picture. This concept is an extension of *Prestel* (which basically uses *alphamosaic graphics*) to use photographic images.

In Picture Prestel, the picture is analysed into *chrominance* and *luminance* components which are digitized and stored as samples. These digitized pictures are stored as a series of *pages* (or *frames*) on the Prestel *database*. When a picture is downloaded into a terminal, it is decoded into 8-bit *PCM* components and stored in a picture memory. A digital-to-analog converter is used to generate the picture display.

This system has been demonstrated to produce high-quality images superior in quality to both alphamosaic and *alphageometric coding* techniques. However, the cost of the *alphaphotographic* terminal and the transmission time required for images imply that it will be several years before the technique can be used in a large public service application.

PIP   ⇨ *path independent protocol*.

pixel   The smallest discrete element which can be referenced in a visual display image. A contraction of the term *picture element*.

plain language transmission   A form of

telegraph transmission in which *signals* are transmitted in accordance with the international standard alphabet and are in the form of *words* which have conventional meanings of the particular language; e.g. English, French.

plesiochronous signals Two or more signals are said to be plesiochronous if their corresponding significant instants occur at the same rate but not in the same *phase*.

point-of-sale terminal (POS) A device situated at a place where commercial transactions take place resulting in cash or credit payments. Such machines are located in shops, garages and stores where contact takes place between staff and customers. They are used to capture details of the transaction and to issue receipts to customers. They may be connected on-line to a credit checking centre, but often they contain a storage device to record transactions throughout the day. They may be polled at close of business by a central *computer* to capture details of the day's business.

point-to-point network A *network* in which all connections between *terminals* and *nodes* are by permanent direct communication over *leased lines*. Such systems are less complex than *switched systems* and speed of response is very high, but they are usually inefficient in the utilization of line plant.

polling A technique used in the operation of a *multidrop network* in which several terminals are connected to only one *line* and to one computer interface *port*. Each terminal can transmit or receive information to and from the *interface* port, but the *computer* decides which terminal is active at any instant. The selection of specific terminals is achieved by *software*, and the process is known as polling.

port A point of entry to a service or *network* and, generally speaking, representing a single *channel* to which a *user* can connect

at a specified *signalling rate*. Thus, when a user to the UK *videotex* service known as *Prestel* dials a connection to the service, he occupies a single port for the duration of the *call* where the port provides *information retrieval* services at a signalling rate of 1200/75 *bits* per second.

POS terminal Abbreviation of *point-of-sale terminal*.

positive justification In *time division multiplexing*, the insertion of *justifying digits* into a stream of *information bits* in order to create coincidence with a required series of *time slots*. ⇨ *justification*.

possible crosstalk *Crosstalk* components which exist but do not intrude on the *user* at the point at which they have been measured but may intrude at another point.

possible crosstalk components Unwanted *signals* which arise from *speech signals* and which are not of sufficient *amplitude* to interfere with desired signals at the point at which they have been measured, but may interfere elsewhere. ⇨ *crosstalk*.

preamble sequence A series of *bits* transmitted by a *station* to a *channel*, to precede each *frame* in order to establish *synchronization* with other stations using the channel. This technique is used in *local area networks*.

prefix Any set of coded information which qualifies information that follows; e.g. an *international dialling prefix* which specifies the country for location of the *subscriber* referenced by an ensuing set of *address digits*.

preparatory period A period of time allocated to maintenance operations prior to transmission of a television broadcast signal over a *circuit*. During this interval, engineers carry out tests and adjustments.

presentation control Those elements of a

system, designed in accordance with the principles of an *open systems architecture*, which deal with the *data* formats and transformations required by end *users*.

Prestel The trade-mark and name given by British Telecom to the *public videotex* service in the UK. Prestel was the world's first service of this type (also known as a *viewdata* service). More information is given under the article entitled *videotex system*.

preventive maintenance Work carried out to identify and clear faults before they can affect the quality of service offered by a system.

primary centre An *exchange* which *interfaces* a number of *local exchanges* to a *trunk network* in a *telephone system*. Also known in the UK as a *primary trunk exchange* and in the USA as a *toll center*.

primary channel A *channel* attached to a device such as a *modem* as the main path for communication, as distinct from a *secondary channel* used for diagnostic purposes.

primary colours The three primary colours of the *visible colour spectrum* are red, green and blue, which can be combined to create other colours which can be perceived by humans. They are, therefore, used in television systems to create a full range of colours for the images to be displayed.

primary routes *Circuits* normally used for a given purpose in a *network*; but *alternative paths* (or *secondary routes*) may exist.

primary trunk exchange An *exchange* which accepts *calls* from *local exchanges* to which it is connected by a *trunk* or *toll circuit*. It may *route* such calls to other *local exchanges* connected to it, or route the call via other *primary*, *secondary* or *tertiary exchanges*.

In Britain, the term *group switching*

centre is used to describe such an exchange, and in the USA *toll center*.

print mechanism A mechanical device used for printing *information* from a *computer* or a *data terminal*.

priority facility A feature of a communications system which enables certain *users*, or classes of *traffic*, to be given preference over others.

private circuit connection A connection made to a *network* using a *local circuit* or *leased line* exclusively allocated to a particular *subscriber*.

private leased circuit A *circuit* reserved for the use of a particular organization to provide communication between two points for either *voice* and/or *data traffic*; the circuit being hired from a *common carrier* (e.g. *PTT*).

private line A *communications circuit* intended exclusively for the use of a particular person or organization and not available for use by others. The connections to the circuit terminate at private locations, and access cannot be obtained ordinarily from a *public network*.

private network Any *network* designed and operated for the exclusive use of a particular organization or group of *users*. It may use some public *circuits* but, in principle, circuits are leased and reserved for exclusive use, and *network management* and *control* functions are operated by, or on behalf of, the organization.

private videotex system A *videotex system* set up by a corporation for internal communication purposes to collect and distribute information over the *public telephone network*. Intended as a means of communication between staff in the organization, it may also communicate with people who are agents of the corporation or with whom the corporation normally does business.

129

private viewdata system ⇨ *private video-tex system.*

probability of call blocking In a *circuit switched system*, the probability that a requested connection may not be made at all. ⇨ *call blocking* and *availability of service.*

probability of excessive delay In a *circuit switched system*, the probability that a *user* will not: (a) receive a *dial tone* within a certain time of requesting a *call*; (b) receive a connection within a certain time of dialling.

probability of failure A measure of the reliability of equipment which is expressed as the probability of the equipment failing over a given period of time; e.g. a failure once every 10,000 operational hours.

procedural model A definition of the functions to be provided in a particular system and specifying the required behaviour of elements in the system.

proceed-to-select An event in the operation of a *call* over a *data network*, in which a *terminal* is advised that its *call request* signal is recognized and *digits* representing the *address* to be selected may now be transmitted.

proceed-to-send signal A *signal* generated by a *network* (e.g. a *dialling tone* signal) when it is ready to receive *calling signals* from a *terminal*.

processing logic The rules designed for operation of a particular *application* or system function and realized either by *hardware* or *software* or a combination of both.

program development The preparation of *computer* programs to carry out *applications* entailing: specifying the procedures required, writing the program of instructions, preparing test data, compiling and generating object code, performing tests to isolate and correct errors.

programming language A language designed to be understood by humans and used to provide instructions to *computers* and computer-controlled devices. Some languages, known as *high level languages*, have the appearance of mathematical notation, or look like conventional language-forms, such as English. Some languages, however, have forms which are convenient to use inside the machine but are not easily learnt by humans. These are known as *low level languages.*

propagation time The time required for a *signal* to travel from one end of a *network* to the other. The *round-trip propagation time* also includes the time to propagate back to the *transmitting station* to detect errors, or *contention* for network resources.

protocol A set of rules defining the way *information* can flow in a system. In all forms of communication, a protocol has to be observed to ensure that correct interaction takes place between *transmitting* and *receiving stations*. In *data networks*, such protocols are formal and often elaborate. They involve a set of agreements (sometimes based upon international standards) which cover:

syntax:	the structure of commands and responses in *bit patterns* and *field* formats
semantics:	the sets of requests, responses and actions which can be performed by each *user*
timing:	the definition of the sequence or ordering of events

Within many modern protocols the concept of *peer interaction* is observed, whereby elements of the system performing a given set of functions at each end are related by a protocol. For example, *addressing* and *routing* functions have a related protocol; and *data link control* elements interact to an agreed protocol to ensure *error detection* and *error correction.*

These protocols effectively define *interfaces* to communication facilities and are exemplified in a series of *CCITT* recommendations (see *V series* and *X series* recommendations) which define electrical and logical connections for *link control* and *data transfer*.

protocol conversion The process entailed in converting the information used to support transmission of a *message*, as it is required by one *protocol*, into the form required by another protocol. Such conversion is achieved by *computer*-based equipment which may have to be designed to achieve an interface between incompatible protocols in a *data network*.

PSE ⇨ *packet switching exchange*.

psophometric weighting function ⇨ *psophometry*.

psophometry The name given to the studies concerned with the effect upon *users* of *noise* interference on a *telephone line*. The different frequencies at which noise may arise create different levels of annoyance to users. As a result of such studies, a weighting function has been given to noise at different frequency levels; this is known as the psophometric weighting function.

PSS Abbreviation of *packet switching service*.

PSTN Abbreviation of *public switched telephone network*.

PTT The legal entity responsible for the regulation and operation of all public communication services in a country. A PTT may license others to operate services on its behalf, and the powers of a PTT may vary from country to country but, in practice, they have extensive regulating powers and often act as communication carriers.

public data networks The first public data networks began commercial service around 1970 and, during the decade up to 1980, such networks were implemented in 30 or more countries. Public data networks are designed to provide low *error rates* and better line utilization using *digital techniques* rather than *analog techniques* which were previously used.

Public data networks have progressed through three distinct stages throughout this period: (a) *digital leased circuits*, (b) *digital switched circuits*, and (c) *packet switching networks*.

Digital leased circuits: In the early stages of this development, *users* wishing to transmit *digital data* over public networks were required to lease circuits from *carriers* to provide a permanent *terminal*-to-terminal connection over a *private line*. However, by using *time division multiplexing*, more than 20 separate *digital data channels* can be transmitted in the *bandwidth* normally devoted to one *voice channel*. Prior to the advent of these digital leased circuits, an analog transmission system normally produced a relationship of one data channel to one voice channel. In a digital network, *data signals* are maintained in digital form, whereas in an analog system the signals are converted to analog form for transmission and reconverted to digital form on arrival at the destination. A major advantage of digital circuits over analog transmission is the ease with which the original signal can be regenerated along the channel to arrive at the destination in its original form. The introduction of digital circuits improved the *error rate* to 1 *bit* in 10^7 bits, compared to 1 in 10^5 as experienced in the use of analog circuits.

Digital switched circuits: The next stage in the evolution of public data networks was the provision of *switched circuits*; i.e. users could dial a connection and be charged only for connect time. At the same time, carriers were able to provide fast *call set-up* and *call clearing* mechanisms. Typically, the *billing* charges in such digital systems can be recorded in one-second increments, thus providing greater efficiency to the end user. The terminal-to-terminal connection is for the duration required, rather than a permanent terminal-to-terminal connection.

Packet switching networks: A further stage of evolution is taking place with the introduction of packet switching networks. Where messages are short and the data volumes are not high, an even greater utilization of the carrier network can be achieved by using packet switching. Essentially, the paths through the network are shared by a number of users at the same time. It is not apparent to the users that this is taking place but, in fact, their individual *message signals* are each divided into little *packets* consisting of from 16 to 1024 bits. Each packet contains sufficient leading information to enable the system to control delivery to the *called terminals*. Packet networks provide a *virtual circuit*; i.e. to a pair of terminals it appears that a point-to-point connection exists. In fact, major paths in the *trunk routing* are also being used momentarily to transmit packets forming parts of other signals. ⇨ general article on *packet switching*.

Each of the forms of transmission described above is found to exist today, and a family of *protocols* known as the *X series interfaces* has been drawn up by the *CCITT* to govern connection to public data networks. An earlier series, known as the *V series interfaces*, was introduced c. 1960 to govern attachment to analog facilities.

public data transmission service A service supplied to the public for the transmission of *data traffic* using a public *data network* operated by a communications authority.

public switched telephone network (PSTN) The term used to describe the public *telephone system*, including the *telephones*, *local lines*, *local exchanges*, and the complete system of *trunks* and the *exchange hierarchy* which makes up the *network*.

public telegraph network A *network* established to provide a *telegraph service* for public use and supplied by a *telecommunications* operating authority.

public videotex system A communications system intended as a public information service, and to provide for an interactive service over the public telephone network to members of the public and the business community, in which information is displayed on an adapted television receiver. ⇨ *videotex system*. Contrast with *private videotex system*.

public viewdata system ⇨ *public videotex system*.

pulse amplitude modulation (PAM) The name given to a technique which forms the basis of *pulse code modulation (PCM)*. The technique consists of *sampling* an analog *speech signal* at specific short intervals to produce a series of pulses of identical duration but varying in *amplitude* in accordance with the original speech waveform; ⇨ Appendix 8. The number of pulses required to represent an *analog signal* accurately depends on the frequency of the analog signal but, to give a good reproduction of speech, 8000 pulses per second are required. In practice, these pulses are modified to a coded form to avoid *distortion* in transmission. ⇨ *pulse code modulation* and *quantization*.

pulse code modulation (PCM) A method of converting an *analog signal* (e.g. a *voice signal*) to a *digital signal*. This process entails *sampling* the analog signal and encoding to represent the signal levels of the waveform of the analog signal in digital form. The sampling technique is based upon a principle known as *pulse amplitude modulation (PAM)*; ⇨ Appendix 8.

In general, the sampling rate is at least twice the frequency present in the analog waveform, and a practical rate for speech is 8000 samples per second. In fact, digital signals of varying *amplitude*, but identical duration, are produced by PAM; in practice, such signals are prone to *distortion* in transmission. Each pulse representing a sample is thus further coded as a *group* of *binary digits* with only two levels of amplitude. A *CCITT* standard recommends that each sample consist of an 8-*bit* code; thus a *speech channel*, after digital encoding, would have a *bit rate* of 64,000 bits per second. PCM techniques are used in association with *time division multiplexing (TDM)* to carry many speech signals along a *transmission system* as interleaved digital signals. At the receiving end, the encoded digital signals are reconverted to analog signals and delivered to *users*. The great advantage of TDM is that it avoids many *crosstalk* limitations which arise with analog signals on *twisted wire pairs*, and it provides a very efficient form of *multiplexing* ⇨ entry on *time division multiplexing*.

pulse stuffing　⇨ *justification*.

punched card　A card, usually measuring 7⅜″ × 3¼″ and ·007″ thick, in which holes are punched to represent *data*. One of the earliest forms of *computer* input, they have been extensively replaced by direct entry systems in which the data entry staff are on-line to a processor which collects and validates data.

punched paper tape　A method for preparing *information* for a slow-speed *communication channel* (e.g. *telex*), in which

information is represented as patterns of holes punched as rows across a narrow strip of paper tape. Various standards exist and paper tape may have 5, 6, 7 or 8 holes in each row. For use in the telegraph service, 5-channel tape is common but, for *computer* input where a larger *character set* is often needed, 7- or 8-channel *paper tape* is used. ⇨ *torn tape centre*.

push button dialling　A method of generating *digit signals* using push-buttons rather than a dial. The pulses are generated either by disconnection of a *circuit* when each button is pressed, or by activation of *oscillators* which send distinctive tones to the line. ⇨ *multifrequency tone signalling* and *loop-disconnect signalling*.

push button telephone (PBT)　A *telephone* in which the *user* is able to obtain an *address* by pushing buttons, rather than by operating a dial. This method may be associated with *loop-disconnect signalling* and *multifrequency tone signalling*.

PWR　Abbreviation for power, used in specifications and on fascia plates of communication devices.

Q

QAM Abbreviation of *quadrative amplitude modulation*.

quadrative amplitude modulation (QAM) A form of *modulation* which uses combined techniques of *phase modulation* and *amplitude modulation*. The system is based upon *signals* of constant frequency, but 4 signal variations are possible by creating a phase difference of 90°. Each of these signals can also be varied in *amplitude* and, for example, if 4 amplitude variations are allowed, the signals can be used to represent up to 16 separate values. Used for the transmission of *binary coded information*.

quality of a circuit The description of a *circuit* is usually given in technical terms which describe its limitations for transmission of *signals* of a particular type. It is difficult to define the precise quality of a circuit without considering the equipment to be attached to it.

Communications authorities generally provide circuit specifications for equipment manufacturers so that they can design equipment to make good use of the circuit characteristics. The communications authorities generally aim to maintain the quality of circuits to conform to these specifications, but in *switched systems* (e.g. the public *telephone network*) the equipment used is selected for each *call* on a random basis. The increasing use of the telephone network for *data communications* has also introduced the need for different levels of specification. The main topics covered in such specifications are listed below and are defined elsewhere in this book: *frequency band, loss/frequency response, group delay/ frequency response, random noise, signal-to-quantizing noise, maximum frequency error, transmit-to-receive crosstalk, signal-to-listener echo ratio, variation of insertion loss with time, error rate*.

quality of service A measure of the diffi-

culties a *user* experiences in using a system as shown by the quality of the *lines* and the extent of *call blocking* that takes place. ⇨ *availability of service* and *reliability of service*.

quantization A process associated with *pulse amplitude modulation (PAM)*, involving the division of an *analog signal* into digital form by *sampling* and the reassembly of the *digital signals* after transmission into an analog waveform. In this process, samples are classified into intervals, each having a single value to represent the *amplitude* of the sample. In a practical system for *voice* transmission, each sample pulse may be presented by 7 *bits*, giving up to 128 *quantization levels* to represent one sample. ⇨ Appendix 8.

quantization distortion The *distortion* which can occur in an *analog signal* from the process known as *quantization*. ⇨ *quantization error*.

quantization distortion power The power of the *distortion* component of an output *signal* from the *quantization* process.

quantization error Relates to *pulse code modulation (PCM)*; an error arising in the reassembly of a *digital signal* to an *analog signal* in the process known as *quantization*. Such an error would arise if too few *quantization levels* were selected; the effect for the *user* would be *distortion* of the resulting analog signal. ⇨ *quantization distortion*.

quantization interval The interval between two adjacent values in the process known as *quantization*. ⇨ *quantization level*.

quantization level In the technique known as *pulse code modulation*, signals representing speech are converted from *analog* to *digital* form for transmission. Samples taken of the *analog signals* result in pulses of varying *amplitude*, which are then each represented by a group of *bits* which describe the amplitude value of the pulse. This process is known as *quantization*. Using 7 bits, up to 128 levels of amplitude can be represented. Thus, such a system has 128 quantization levels.

quantization noise   An interference which arises during the process of *pulse code modulation* in which *noise* is generated during the conversion of *analog signals* to *digital signals* and vice versa.

quantizing   ⇨ *quantization*.

quarternary trunk exchange   One of a set of *trunk exchanges* providing a means of communication in a *telephone network* at the fourth level. In an *exchange hierarchy* they would be fully interconnected to one another and each will represent the convergence of lower levels of the hierarchy including *tertiary*, *secondary*, *primary* and *local exchanges*. ⇨ Appendix 12.

quiescent phase   The non-active phase during the establishment of a *call* using a *switched* or *leased circuit*, in which the *network* and the *called terminal* indicate the ready or not-ready status.

R

radar A system that detects distant objects by measuring the deflection of *microwave* frequencies, used, for example, to provide information about the movement of aircraft, ships or vehicles.

radio frequencies The spectrum of frequencies used for *radio transmission* and generally considered to be within the range from 20 kHz to 3 GHz. ⇨ article on *electromagnetic waves*.

radio frequency carrier wave A *radio frequency signal* used to carry another signal imprinted upon it by the techniques of *modulation*.

radio frequency signal Any signal falling within the *radio frequency spectrum* (20 kHz to 3 GHz) and used to radiate information through *space*. ⇨ article on *electromagnetic waves*.

radio frequency spectrum The range of *electromagnetic waves* associated with radio and television broadcasting, including transmission in the following *bandwidths*:

low frequency	
(*LF*)	30 kHz to 300 kHz
medium frequency	
(*MF*)	300 kHz to 3 MHz
high frequency (*HF*)	3 MHz to 30 MHz
very high frequency	30 MHz to 300
(*VHF*)	MHz
ultra high frequency	
(*UHF*)	300 MHz to 3 GHz

This spectrum covers the present range of frequencies used for commercial and military/civil communications.

radio receiver A device which can be tuned to receive *radio frequency signals* and to detect *sound signals* carried in the broadcast *radio signal* for audible presentation to listeners.

radio signal Same as *radio frequency signal*.

radio transmission A form of *telecommunication* in which *message signals* are radiated as energy in the form of *radio waves*. Examples of radio transmission include sound and television public *broadcast* systems, established to serve a particular geographic area, as well as directional radio systems designed to provide a *channel* for point-to-point communication.

radio transmitter A device capable of capturing *signals* representing *sound* or light and converting them by a process of *modulation* into a form suitable for transmission as *electromagnetic waves* in the *radio frequency spectrum*.

radio waves A range of *electromagnetic waves*, in the *frequency spectrum* from 20 kHz to 3 GHz, which are suitable for carrying *signals* across *space*.

radiometry The measurement of light energy from a source, irrespective of human factors involving the perception of light. Studies concerned with consideration of the human factors are known as *photometry*.

RAM ⇨ *random access memory*.

random access memory (RAM) A type of storage device used in *computers* to retrieve information directly with very short access times measured in milliseconds. Any particular retrieval operation will produce information within a more or less constant access time, regardless of the last location addressed. Magnetic disc storage is an example of such a storage medium, but *magnetic tape* is not a random access memory medium. ⇨ *main memory*.

random noise A form of interference which may be noticed in a *telephone channel* as a hissing sound. If the noise level is high in a particular *circuit*, it can severely impair *signal* quality at the *receiving terminal*. *Noise* is measured in *decibels* and expressed as a ratio of noise power to the power of a test signal. A most important ratio is the power of the random noise relative to signal power (the *signal-to-noise ratio*). Also known as *white noise* or *gaussion noise*.

RAP ⇨ *remote access point*.

raster The pattern formed by the movement of an *electron beam* inside a television camera tube (or *TV receiver* tube), to scan a picture and form a *signal* relating to the light emanating from objects being televised. ⇨ *video signal*.

RBT Abbreviation of *remote batch printer*.

read only memory (ROM) A storage device, the information contained within which is fixed during a special process and cannot be overwritten by a *user* of the device, or by any *computer* program during the normal operation of the device. Usually used to build programs such as *compilers* into a *microprocessor* or *terminal*.

ready-for-data signal A control *signal* transmitted by a *modem* to a *data terminal*, to indicate that a connection is available to allow *data* to be transferred to or from another data terminal. The ready-for-data condition in a *data network* occurs when all connections involved in the *call* have been completed.

ready for sending An *interchange signal* which passes from a *modem* to its associated *data terminal* (*DTE*) to indicate that the modem has established contact with a distant modem and that *data transmission* from the DTE to the distant location can take place. ⇨ *request to send* and *carrier detector*.

ready state A condition in which a *data terminal* (*DTE*) and its *modem* (or *DCE*) are operable, are not engaged in any activity associated with a *call*, and the DCE is ready to accept an *incoming call* or a *call request* signal from the DTE.

ready to send ⇨ *ready-for-sending*.

real-time system A special class of *on-line system* in which the processing of transactions occurs simultaneously with the events which give rise to transactions. The relationship between the various *users* and the *computer* system is such that business cannot be conducted at all without the system being available, and any user who accesses the *database* will always find an up-to-the-second picture of the events and objects being controlled by the system. An airline ticket reservation system is an example of a real-time application.

receive only (RO) Pertaining to a *terminal* device which can receive *information* but not transmit information. It might, for example, be a *data printer* or a *paper tape punch*.

receiver clock In *synchronous transmission*, a device which takes its timing from line conditions, and establishes the *time intervals* within a *receiving terminal* to keep the terminal in *synchronization* with signal pulses sent by the *transmitting station*. ⇨ *transmitter clock*.

receiving aerial A device designed to receive *radio waves* to be input to a *radio receiver* where selection and *amplification* of specific frequencies may take place.

receiving earth station A system forming part of a communications *network*, and equipped with a dish-shaped *aerial* to receive *radio signals* reflected from a *satellite* orbiting the earth. The *earth station* may distribute the received signals over an *earth network* or pass them to another *transmitting earth station* to be beamed at another orbiting *communications satellite*.

receiving station Same as *receiving terminal*.

receiving terminal A *terminal* receiving *signals* from another, or from a network *node* at a particular instant.

record Any unit of *data* representing a complete item of information, or a transaction to be processed by a *computer* system and consisting of a defined set of related *fields* of *information*.

record separator (RS) A special *code* under the general classification known as *information separators* and used in a *data communication code* to denote the boundary between *records* in *binary coded* form.

recorded information service A service which allows members of the public to dial into a special *terminal* using the *telephone network* to receive pre-recorded information; e.g. sports news or a speaking clock.

recovery procedures Procedures, usually implemented by *software*, which are designed to protect an *on-line* or *real-time* *system* from major failure. Should any component of the system fail, potentially causing a major interruption of service to the *users*, the recovery procedures will automatically record the status of current transactions, and will write away to *backing store* the contents of any *main memory* locations. Any reserve equipment may be brought into operation automatically and faulty units be identified and isolated for the attention of system engineers. The system may then recover automatically by retrieving the information previously stored.

The complexity and sophistication of such recovery systems depends upon the importance of the *application* to the end users, but in any real-time *data processing* application, or in a communications system, it is expected that large sums would be invested to prevent a major failure, and automatic recovery would be effective in seconds or minutes at the most. This would

entail the employment of redundant *hardware* units ready to be brought into service by the operating system, and the ongoing maintenance of duplicate file systems and *transaction* records.

Recovery systems must be prepared for the possibility of failure by frequently taking copies of *files* and *records*, and maintaining a *transaction log*, so that, in the event of a serious failure, the recovery routines can reconstitute the status of files quickly and efficiently without loss of *data* and without requiring the users to repeat previous work.

Red, Green, Blue (RGB) Relating to the primary inputs which are basic to all colour television systems. The light from a televised image is separated into three separate beams by *filters* before passing to one of three camera tubes representing the red, green and blue components of the light from the image. The separate *signals* are then *modulated* on to a *carrier wave* and transmitted as separate components within the overall television signal. The receiver recovers these separate signals after *demodulation* of the carrier wave and combines the signals to re-form the colour image from the original scene.

redundancy A concept used in the design of communications systems in which more functional units are provided in the system than are strictly needed to handle the planned workload. The units are arranged such that another unit is easily and automatically substituted for any unit which fails. Several parts of the system may fail, but the system continues to operate by falling back on spare, or lightly loaded, units.

redundancy checking A method of *error detection* in *data transmission* in which additional *bits* or *characters* are computed from the *data*, and are added to *blocks* of data transmitted over a *network*. The *receiving terminal* is able to carry out checks by comparing *bit patterns* formed in the data blocks, in such a manner that it is possible to detect bits which have been lost or incorrectly received. The receiving terminal usually requests automatic retransmission of *error blocks*; but ➪ *forward error correction* and *cyclic redundancy checking*.

reed relay A device used to provide a *crosspoint* in a *circuit switching exchange*. Reed relays are cheap and reasonably compact devices, and they can be combined to make up a switching matrix of any desired capacity. A reed relay consists of a glass tube surrounded by a coil. Within the tube are two nickel/iron reeds with gold-plated tips. When a current is passed through the coil, the reeds are magnetized and come together to form a contact.

In more modern *digital exchanges*, reed relays are replaced entirely by *electronic crosspoints*.

reference carrier A *signal* transmitted over a *multiplexed channel* to provide a stable reference point for detecting *message signals* transmitted upon *carrier waves* at adjacent frequencies. Same as *reference wave*.

reference wave A *signal* used to detect *message signals* which have been generated by the process of *modulation*. In the process of *demodulation*, it is necessary to use a replica of a *carrier signal* to detect the original message signal. Sometimes the carrier is generated locally in the *demodulator*, but in other cases (particularly where a large number of *channels* are created over a suitable physical path) one channel is used to transmit a *reference carrier* so that other carriers can be generated from it to demodulate signals on the various channels. Same as *reference carrier*.

regenerative repeater A *repeater* which operates in *transmission systems* using binary *digital signals*. It examines signals and identifies each element as 0 or 1, and automatically regenerates appropriate 0 or 1 elements for output to the *line*. It has the advantage of generating virtually noise-free signals which are exactly synchronized

with the pulses of the incoming *message signal*.

regeneration unit A device used in *digital transmission* to reconstitute a distorted *signal* as it passes along the *transmission path*.

regional exchange A *node* in a communications *network* which controls *traffic* between different geographic regions. For example, in the British *telephone network* each region comprises several areas, and each area exchange controls several *local exchanges*. The regional exchange thus controls *trunk circuits* providing major links in the overall network. ➪ *exchange hierarchy*.

register
1. A *control* device used to receive and store *address information* during the set-up of a *call*.
2. In any *computer*-controlled system, a special store location having special properties used for arithmetic or logical operations.

register translators Units incorporated into *strowger* telephone exchanges to assist *users* by simplifying the dialling process. ➪ *director exchange*.

reliability of service A measure of the extent to which *users* can rely on the service as shown by the effective operation of *terminals* and transmission equipment. The reliability of a communications system is generally expressed as a set of target *probabilities of failure* for various parts of the *network*. For example, in a *telephone network*, target probabilities for failure per annum might be: telephone terminal 0.1, *local line* 0.075, *local exchange switching equipment* 0.01, and so on for other system components. ➪ *availability of service* and *quality of service*.

reliability target In order to plan for an acceptable level of satisfaction for *users* of a communications system, it is customary to establish reliability targets for the system and for key system components. The targets are expressed as the *probability of failure* per unit time. For example, the reliability target for a *node* in a *network* might be 0·02 per annum. With 50 such nodes in a network, the expectation is that, in one year, 50 × 0·02 (or 1) failure would occur. ⇨ *mean down time* and *fault rate*.

remote access point (RAP) A *multiplexor* or *concentrator* established in a particular location to provide access to a *network* in areas which are not directly covered by *nodes* having full *network facilities*. The RAP will normally be connected to a main node by a high-speed *channel*, capable of carrying a volume of *multiplexed traffic* to the *main network*.

remote batch printer (RBT) A printer connected as a remote *terminal* under control of a central *computer* and used to print *information* in batches after the information has been prepared by the computer.

remote batch terminal A *terminal* used to collect *data* which is entered from a *keyboard* or some other local data input device. The data is stored on a local storage medium (e.g. a magnetic disc) and is later transferred as a complete batch of *records* over a *transmission channel* to a central computer installation. Such a terminal may also receive batches of data from a central *computer*.

remote electronic blackboard A terminal device used in distance teaching systems, in which members of a class are joined together in a seminar or conference using a communications network. Each user is equipped with a visual display and/or graphic tablet. By writing on the screen with a light-pen, or on the graphic tablet, each user is able to generate graphic images which can be seen and amended by others participating in the conference. ⇨ *teleconferencing*.

remote printer A mechanism used for printing *information* at a location remote from the centre at which the information is stored.

remote printing The production of printed output in a communications system at a point remote from the location at which the *information* is stored or entered into the system.

repeater A type of *amplifier*, used at regular intervals along a *transmission channel* to regenerate *message signals* and overcome various impairments, such as *attenuation* and *crosstalk*. ⇨ *analog repeater* and *digital repeater*.

replicated database A computer networking *application* in which duplicate copies of a *database* are transmitted to be available at regional centres for local enquiry. The updating of such *files* is usually done centrally to simplify control problems, and corrected versions of the database (or parts of it) are transferred periodically to the regional centres. Such designs are best used with *data* which is relatively slow to change, and where the operational requirements of *users* do not require an up-to-the-minute view of the information. Contrast with *partitioned database*.

request-response unit In a *packet switching network*, this term refers to the *application* information being transferred between *users*, as part of a *packet* which is, in turn, part of a *transaction* between the users. The packet itself is completed by the addition of *header information*; e.g. for *data link control*, *path control*, and *transmission control* within the *network*.

request-to-send An *interchange signal* which passes from a *data terminal (DTE)* to its *DCE*, indicating that it wishes the DCE to transmit information to the *line*. The DCE must reply with a *ready-for-sending* signal, indicating that it has established a connection with the DCE at the distant terminal. The data terminal cannot send *data* to its DCE until this condition is established. ⇨ *carrier detector.*

request-unit ⇨ *request-response unit.*

reserve link A *circuit* used as a standby for a regular *link* which normally carries the *traffic* for some designated purpose.

residential terminal A *terminal* used in the home rather than in a business environment. The implication is that a residential terminal will originate and accept a lower *traffic volume* than a business terminal.

resolution A term denoting the degree of detail which can be created in a particular *visual display* system; e.g. a television system or a visual display used to form *text*, *data* or graphical images. The degree of resolution is indicated by the number of *picture elements* which make up the image. For example:

system	matrix of picture elements
USA 525-line TV	340 × 422
British 625-line TV	402 × 572
French 819-line TV	527 × 786

The term *high resolution display* is often used in relation to information systems, but this is a relative expression which implies no specific degree of detail. Factors affecting human perception of images include the nature and shape of images to be drawn, the size of the viewing screen, whether colour or monochrome is desirable, the viewing distance, and the distance between picture elements. ⇨ *visual acuity* and *kell factor.*

response frames A *page* of *information*, designed by an *information provider* for display on a *visual display* terminal, to act as an order form, or some such document, and allowing a *user* of the system to enter details for the ordering or reservation of goods and services. ⇨ *videotex system.*

response header A unit of *data* attached to a *packet* so that it may be identified to

the *transmission control* and *data link control* procedures in a *packet switching network.*

response-unit ⇨ *request-response unit.*

restart A term used to describe the processes involved in recommencing the operation of a *data processing* system (particularly one which uses *data communication* techniques) after a system failure. This often creates a *peak load* on the system because *messages* and *transactions* which have previously been interrupted are waiting in queues to be serviced.

return path A *channel* used to convey *information* back to a *transmitting terminal* to provide *control information* about a *call* in progress. For example, to send information about the status of the *receiving station*, or to send back a copy of the transmitted *message signal* to enable the accuracy of the transmission to be verified.

revertive error checking A system of *error checking* in which the *transmitting terminal* generates a sequence of *check digits* derived from an algorithm performed on the content of a *frame* to be transmitted. The *receiving terminal* uses the same procedure to check the incoming information, including the check digits, and reverts to the transmitting terminal, using the same *channel*, to request a retransmission if an error has been detected. Contrast with *nonrevertive checking.*

rf carrier wave ⇨ *radio frequency carrier wave.*

rf filters Devices intended to accept specified frequencies in a *radio frequency* bandwidth and to reject frequencies falling outside the specified *bandwidth*; e.g. to provide a means of tuning reception to a particular *signal.*

RGB Abbreviation of red, green, blue, the three primary colours which are fundamental to colour television systems. During the transmission of picture information, the light from a televised scene is separated into three outputs corresponding to red, green and blue. These outputs are transmitted upon a *carrier wave* to the receivers. Each TV colour receiver has patterns of red, green and blue phosphor dots upon the surface of the viewing screen. The *demodulated* signals are used to drive separate *RGB guns* which control the direction of beams on to the appropriate colour dots on the screen. The illuminated dots on the screen form combinations of primary colours to reproduce the colour of the original televised scene.

RGB guns *Electron guns* used in *colour television receivers* to activate coloured phosphors on a television screen.

ring network A *network* in which the various *nodes* are interconnected along a *transmission link* which can be diagrammatically represented as the circumference of a wheel. The nodes appear as points along the circumference, and communication between two nodes must proceed via any intermediate nodes along the ring. Also known as *loop network*. Contrast with *mesh-network*, and ⇨ *network topologies*.

ringing tone The *signal* heard by the *user* of a *telephone* when the called telephone is being rung.

RO Abbreviation of *receive only*.

roll-call polling A system of *polling* in which each *terminal* device is addressed in strict sequence by the *controller* to see which, if any, device is ready to send or receive a *frame* of *information*, starting at the nearest terminal and working out towards the most distant. ⇨ *hub polling*.

ROM ⇨ *read only memory*.

round-trip propagation time ⇨ *propagation time*.

router In a *message switching system*, that part of a *node* (or *exchange*) which examines incoming *messages*, interprets the *address information* contained in each message and determines which of the outgoing *links* are to be used. The router is usually a *computer* program which selects messages from *incoming buffers* and places them into appropriate outgoing *message queues*.

routing The operation which takes place to connect two *users* in response to the *addressing information* contained in a transmission request. The combination of *transmission paths* used for a particular *call* is also known as the routing. ⇨ *basic routing* and *alternative routing*.

routing rules These are rules used in times of high *traffic* to determine when, and in what sequence, *alternative routes* are attempted.

RS ⇨ *record separator*.

S

sampling A process in which the value of a variable factor (e.g. a current of fluctuating frequency) is examined at periodic intervals, and in which the variable may be further represented by a system of coding, derived from the samples. ⇨ *pulse amplitude modulation.*

satellite ⇨ *communications satellite.*

satellite channel A *transmission path* provided by an earth satellite system. ⇨ *communications satellite.*

satellite station A transmitting and receiving centre, capable of communicating with an artificial body placed into earth orbit for the purpose of long-distance communication. A highly directional *antenna* is used to control precisely the transmission of *radio frequency signals* to and from the *satellite.*

scanning beam In a television camera, a beam of electrons which is passed over an electrically charged surface to create an electrical current which represents light emanating from a televised scene.

SCL ⇨ *supervisory control language.*

screened cable A cable in which the *conductor* used to carry *message signals* is encased in a sleeve of which the outer layer is an earthed metal skin. The purpose of the screen is to reduce the background noise arising from electrical interference along the cable.

SCU Abbreviation of *system control signal unit.*

SD Abbreviation of *space division.*

SDLC Abbreviation of *synchronous data link control.*

SECAM Acronym for sequential couleur à mémoire. A colour broadcasting system developed in France, but used also in the Soviet Union. It is distinctive in that the two colour difference signals are transmitted sequentially rather than simultaneously, as is the case with the *PAL* and *NTSC* systems. SECAM uses 625 picture lines and a 50 Hz *field frequency.* ⇨ *PAL* and *NTSC.* .

secondary centre An exchange in a *public switched telephone network* which links the *trunk circuits* arising from several *regional exchanges* to the main *tertiary trunk exchange* network. ⇨ *exchange hierarchy.*

secondary channel A *channel* attached to a device, such as a *modem,* to enable tests and diagnostic information to be obtained about the performance of the modem without interrupting the *primary channel,* which is used for *data transmission.*

secondary characters A set of special *characters* occurring in the *figures shift* of a *telegraph code* including, for example, punctuation symbols and % @ £.

secondary route An alternative path for transmission of a *message* between two *terminals* over a *network,* used when the *primary route* is unavailable due to congestion or maintenance.

secondary trunk exchange An *exchange* which handles major *routings* in a public *telephone network* and which is connected only to other *trunk exchanges.* ⇨ general article on *exchange hierarchy.* Also known as *secondary centre,* and in Britain as *district switching centre.*

selection digits A group of signal elements representing the *digits* of a telephone *number* of a *called party.*

selection signals A sequence of *characters* which appear in a *facility request* and which define the *address* required for a *call* and also the particular *network facilities* needed in making the call.

selector An electromechanical switching device used in the old *strowger* exchanges, in which moving contacts are used to create connections for switching *lines*.

self test A test undertaken by an item of line terminating equipment such as a *modem*, in which the modem is instructed to establish a *loop back* path through itself. This instruction is normally conveyed to the modem by a specialized *network diagnostic controller* which transmits a test pattern through the modem and analyses the returning pattern to assess the performance of the modem under various conditions.

semaphore An early method for telegraphing *signals* between locations in visual contact, aided by the telescope or binoculars. The physical positioning of arms, flags or spars was used to represent information. Later inventions included the optical semaphore (attributed to Claude Chappe, a Frenchman), in which coded information was transmitted as light signals. These developments led to the later introduction of the *electrical telegraph* by Samuel Morse, (⇨ *Morse, Samuel*), around 1830.

send channel Most communication *circuits* consist of two *channels*, to allow simultaneous two-way transmission between *terminals*. The send channel is used for transmission in a particular direction at any instant, the other channel being known as the return or *backward channel*, and being mainly used for *error control* and supervisory purposes.

sender A device in an *exchange* which controls the transmission of *signals* to and from a *line*.

separate channel signalling A system of *signalling* in which a dedicated *circuit* is used to carry *control* signals related to a large group of circuits carrying *message signals*.

Same meaning as *centralized control signalling* and *common channel signalling*.

serial attribute coding A system of coding in which the attributes of a *character* to be displayed (e.g. its colour, height, or flashing/steady indicator) are stored as non-displayable *codes* in the *page store* serially with characters to be displayed. This means that the attribute characters must appear as spaces in the *text* at the beginning of each new sequence although, in practice, this is not a serious problem because it is usual to change attributes at the beginning of a *word* or phrase where the space is needed anyway. This system is not quite so flexible as the *parallel attribute coding* methods, but the method requires less storage in each terminal and is therefore cheap. The first public trials of *videotex systems* have all used serial attribute coding.

serial bit-stream A series of *binary digits* occurring in sequence one after the other, in which the pattern of *bits* within a particular sequence may have a specific meaning.

serial mode Relating to the transfer or transmission of coded information in which the signal elements (*bits*) pass one after another along a single path. Contrast with *parallel mode*, in which separate paths are provided for each bit forming a *character* or *byte*.

serial transmission The most common sequence of *data transmission* in which pulses representing *bits* are sent one after another along the *transmission line*. Contrast with *parallel transmission*.

session A period during which an *access path* is maintained between a *transmitting* and a *receiving station* to allow a series of related *transactions* to be progressed.

session control layer In a system designed in accordance with a *layered architecture* (⇨ *ISO reference model for open systems architecture*), the session control layer establishes, maintains and terminates logical connections for the transfer of *data* between end *users*.

shadow mask tube A type of *cathode ray tube* used in the construction of *colour television receivers* in which a metal mask containing holes is placed between the *RGB guns* and the groups of three coloured phosphor dots known as *triads* on the screen surface. The mask is constructed so that the red gun can only activate red phosphor dots, the blue gun the blue phosphors, and the green gun the green phosphors.

shannon A shannon is a unit in *information theory* which serves to measure the information conveyed by the occurrence of a symbol. It is defined as the negative of the logarithm of the probability that this symbol will be emitted. Logarithms to the base 2 are used.

If logarithms to the base 10 are used, the unit is called a *hartley*.

shannon's Law A method for stating the capacity of a communication line to account for *bandwidth*, and *signal-to-noise ratio*. The law asserts that the maximum capacity in bits-per-second is derived by

$$W \lg (1 + SN)$$

where W = bandwidth
$\lg$ = logarithm to base 2
SN = signal-to-noise ratio.

shared database A *database* used in a *computer* networking environment which can be updated and accessed simultaneously by *users* in different locations, using programs which provide updating and *control procedures* operating independently of the location of the *data*. For example, *records* may be stored in regional centres

which have a primary responsibility for designated records, but the system may have to cater for changes to records related to a region by *transactions* arising outside the region. Such systems are also known as *distributed systems* and have particularly challenging control problems, which have to do with the orderly management of communications *traffic* in the *network*.

shared service In a *telephone system*, this term describes the condition that exists when two *users* are sharing the same *local line*. This arrangement is sometimes necessary to economize in the use of local lines. The system is usually applied to residential users who have low volume utilization, and special arrangements are made to ring the bell of each *terminal* independently and to measure *outgoing calls* for *billing* purposes.

SHF Abbreviation of *superhigh frequency*.

shift character A *character* in a *data communication code* which is used to designate a following group of *codes* in a transmission sequence as having a specific significance. These codes retain that significance until another shift character occurs to revert to the original case; e.g. *letters shift* and *figures shift* used in *International Alphabet No. 2*.

shift in (SI) A *character* in a *data communications code* which is used to indicate that codes which follow it have the meaning specified in the standard code set. Contrast with *shift out*.

shift out (SO) A *character* in a *data communications code* which is used to depart from the standard set and denotes that all following characters have a meaning other than the standard set until a *shift in* code is detected.

short haul modem A device used to connect a *terminal* to a communication *circuit* in which the *signal* frequencies to the *line* are suitable for transmission over only relatively short distances. Used mainly for

145

in-plant systems, rather than for connection to a *public network*.

SI ⇨ *shift in.*

sideband A *signal* which arises when a *radio frequency carrier signal* is *modulated* by a second signal in order to carry the second signal in a *transmission system*. Sidebands are generated by the process of *modulation* above and below the frequency of the *carrier signal* and are displaced from the frequency of the carrier according to the frequency of the modulating signal. They are known as *upper sidebands* and *lower sidebands*.

signal
   1. In *telephony* and in other public networks, an instruction forming part of the process in setting up, maintaining or clearing a *call*. ⇨, for example, *signalling* and *signals*.
   2. In the general case, any electrical pulses transmitted in a network to represent *message* information, or *control information* in handling the process of communication.

signal bandwidth The range of frequencies required to convey a particular *message signal* accurately over a *channel*. For example, a *speech channel* allows up to 4 kHz for transmission of *electrical signals* representing voices, whereas a *television channel* may require up to 6 MHz to contain all the information necessary in a *video signal*.

signal element A discrete pulse forming part of a *digital signal* and having a value determined by the *pulse amplitude*.

signal message A *message* made up of a number of *signal units* and transmitted on a *signalling* channel to control the set-up, maintenance or termination of a *call* in a *data network*.

signal phase ⇨ *phase.*

signal power A measure of the strength of any *electrical signal* and expressed in *Bels* or *decibels* (*dB*).

signal-to-listener echo ratio The ratio of *signal power* to the power of *echo* signals reflected back to the *transmitting station* and caused by changes in the electrical characteristics of a *circuit*. A form of signal impairment in transmission over long-distance *transmission paths*; e.g. intercontinental *telephone circuits*.

signal-to-noise ratio In any *radio transmission*, or in any communication *link*, a certain amount of *noise* is generated, and is carried as background interference to the desired *message signal*. If the signal-to-noise ratio is high, then the message is unlikely to be impaired; if it is low, the signal may well be severely impaired.
   The problem is accelerated by the use of *amplifiers* which will amplify both noise and signal. *Repeaters* containing amplifiers are spaced at intervals along transmission cables to maintain an appropriate signal-to-noise ratio. These amplifiers are designed to amplify signals occurring in a critical frequency *bandwidth* in which the message signal arises; the bandwidth of the amplifier is restricted so that noise outside the critical bandwidth is rejected.

signal-to-quantizing noise ratio *Quantizing noise* occurs in *pulse code modulation* (*PCM*) systems and arises as part of the process of converting an *audio signal* to digital form and vice versa. If the *signal-to-noise ratio* is high, the *message signal* is not seriously impaired, but if the ratio is low the signal may be impaired.

signal unit (SU) A unit of information in a *signalling system* for a *public data network* and consisting of a defined number of *bits* which provide *information* to control the progress of *calls*. ⇨ *acknowledgement signal unit, initial signal unit, lone signal unit, multi-block synchronization unit, subsequent signal unit, synchronization* *signal unit*, and *system control signal unit*.

signalling This term is given to the procedures concerned with the establishment, maintenance and termination of *calls* in a *network*. *Signalling systems* have been evolved to cater for the development of all classes of communication systems, including *telephony*, *telegraphy* and *data communication*.

The simplest form of signalling occurs in a *local telephone network* in which various signals are transferred between the telephone instrument and the *local exchange*. For example, when a *subscriber* lifts the *handset* from the *switch hook*, an automatic *call request signal* is sent to the *line*. The response from the exchange is a *proceed-to-send signal* which is recognizable as a *dial tone*. Individual *address digits* dialled by the subscriber are also signals and other signals include: *called terminal free* – ringing tone; *called terminal answered* – ringing tone stops; *clear* – a party replaces handset. This illustration is over-simplified, and the signalling process includes complexities concerned with *multi-exchange connections* and *inter-exchange signalling*.

The methods of effecting signalling vary considerably with the evolution of *switching* technology, and in any particular *telephone network* various forms may be encountered.

Direct current signalling is dependent upon a *DC* path being created for signals through the *transmission path* used for each call. It has tended to be replaced by *alternating current* signalling in which different signals are represented by *AC* tones applied to the transmission path. This form of signalling can itself be classified into *in-band signalling* and *out-band signalling*.

In the former case, the tones are conveyed within the *bandwidth* allocated to each *speech channel*, and in the latter case, conveyed in a gap of 900 Hz between speech channels. Signals can be distinctly conveyed as specific tones, and the duration of the tone can also serve to define the specific meaning.

It has been usual for signalling to be carried on the actual *transmission channel* or on a subsidiary channel directly associated with it. Various names are given to this technique, including: *associated channel signalling*, *channel associated signalling*, *decentralized signalling*.

With the use of modern digital techniques, specific signalling channels are designated to control a large number of transmission paths. This is particularly used between exchanges. Terms applied to this technique include: *common channel signalling*, *centralized control signalling*, *separate channel signalling*, *non-associated signalling*. In such systems, each signal is represented by a group of *bits* which include the *channel identification* and the specific signal intended. The exchanges concerned in such a system are able to carry out digital processing and are connected by a *digital data link*.

An example is the *CCITT* system, recommended for *international exchanges*, known as *signalling system No. 6*.

signalling interworking A concept which arises when an existing *network* has to be extended or gradually replaced by a more modern network which uses different technology and therefore different *signalling* standards. There is a need during the period of change-over to provide subsystems capable of recognizing the new and old *signalling systems*.

signalling rate The rate at which a device can transmit or receive *information*, usually expressed in *bits* per second. For example, 1200 bits per second implies the ability to send/receive 1200 pulses per second.

signalling system No. 6 A *common channel signalling* system defined by the *CCITT* for use on international *networks*, in which each *signal* is represented by 28 *bits* including the signal itself, the *channel identification* and *error control* bits. Exchanges operating to this method have computer-like facilities and are linked by a *data channel* over which signals are exchanged as *messages*. ⇨ general article on *signalling*.

signals *Information* transmitted within a *network* to control the handling of *messages* and the set-up, clearing and·maintenance of connections. The signals listed below are defined in this book. The principles concerned are described under the general article entitled *signalling*.

access-barred·signal
address-complete signal
address-incomplete signal
address signal
answer signal
blocking signal
call accepted signal
call answered signal
call connected signal
call failure signal
call not accepted signal
call progress signal
call request signal
called terminal answered signal
called terminal engaged signal
called terminal free signal
calling indicator signal
changed number signal
character signal
circuit group congestion signal
clear back signal
clear confirmation signal
clear forward signal
clear request signal
confusion signal
correct signal
continuity-failure signal
data transfer requested signal
distributed frame alignment signal
end-of-address signal
end-of-block signal
end-of-pulsing signal
frame alignment signal
hang-up signal
line-out-of-service signal
no charge answer signal
octet timing signal
out-of-order signal
proceed-to-send signal
ready-for-data signal

These and other signals are used variously in *telegraphy*, *telephony* and in *data communications*. There are standard systems of *signalling*, and particular reference should be made to a *CCITT* standard for the appreciation of such systems. There are also variations of international standards used in national networks, and in most countries one finds more advanced systems having to operate alongside older, and even outdated, *signalling systems*.

silicon chip A very small device composed of silicon and used for the construction of micro-electronic circuits in electronic machines. Silicon materials are very cheap; their main property is that they are semiconductors and can be used to provide very high-speed *switching* operations necessary in digital storage and processing systems.

simplex
1. Sometimes used to indicate a transmission *circuit* in which *messages* can be sent only in one specific direction.
2. The *CCITT* definition states that a simplex circuit allows transmission in either direction, but only one way at a time. But see definition (1) of *half duplex*.

single bit error An error occurring in a *data transmission* sequence in which a single *bit* is inverted; i.e. a 0 becomes a 1 or a 1 becomes a 0. ⇨ *transmission errors*.

single current circuit A *circuit* used for *data transmission* in which voltages are applied directly to a *line* to represent *binary numbers*; i.e. positive voltage to represent 1 and zero voltage to represent 0. Contrast this with a *double current circuit* in which 1 and 0 are represented by positive and negative voltages respectively. Either system can be used for transmission over short distances (a few kilometres), but for long-distance communication over a *network*, a *modem* or *network interface unit* is needed. Such circuits are often used between a *data terminal* (*DTE*) and its *DCE*, when they are referred to as *interchange circuits*.

single sideband modulation A *carrier wave* modulated at a single frequency produces three simultaneous *signals*: the unmodulated *carrier* and two associated steady frequencies spaced on either side of the carrier by the frequency of *modulation*. If a carrier wave is modulated by a signal which is not steady (e.g. an *audio signal*), a *spectrum envelope* of associated frequencies known as a *sideband* is generated. To *broadcast* and receive accurately *audio frequencies* up to 15 kHz, the sidebands will spread 15 kHz on each side of the carrier wave.

These sidebands are known as *upper sidebands* and *lower sidebands*. It is common practice to select one of these sidebands when transmitting modulated signals over a *transmission channel*, and to suppress the carrier and the other sideband. This economizes in the use of power and *bandwidth* in transmission. ⇨ *multiplexing*.

sinusoidal Pertaining to a *signal* which is in the form of a pure sine wave and in which there is one complete reversal of phase in a cycle of given duration.

SITA Abbreviation of *Société Internationale de Télécommunication Aéronautique*.

sky-waves Radio broadcast signals which are reflected from the *ionosphere*, usually in the *medium frequency* and *high frequency* bands.

SNA Abbreviation of *systems network architecture*.

SO ⇨ *shift out*.

Société Internationale de Télécommunication Aéronautique An international body which has introduced standards for a communication system which is the *main network* used by world airlines for *message switching*. A number of computer *nodes* are installed in major cities and these service *terminals* in airline offices, airports and centres of administration.

soft copy A *message* produced in a communication system but not in physical form; e.g. a message displayed on a *visual display*. Also known as *transient copy*. Contrast with *hard copy* or *permanent copy*.

software Relating to any logical rules which are built into a device by *computer* programs rather than by the physical arrangement of *hardware* components. The scope and application of the software varies greatly. It can be the *operating system* provided by the computer manufacturer or an *application* written by a *user* for a specific purpose.

software interface The specification of the logical rules which will allow two *computer* programs to interact with each other, and to ensure the accurate and efficient transfer of information between the programs.

SOH ⇨ *start of heading*.

solenoid A type of switch, the operation of which depends upon the principles of electro-magnetism. A current passes through a coil which is constructed around an armature which moves under the influence of the current to open or close a contact.

SOM ⇨ *start of message*.

sound ⇨ *sound waves*.

sound bandwidth Relating to the range of frequencies occupied, by *electrical signals* corresponding to *sound waves* and generally considered to be within the *band* from 20 Hz to 20 kHz.

sound carrier A *carrier wave* used to carry an *audio signal* and used particularly to distinguish from a *vision carrier* in the composite signal used for television broadcasting. ⇨ *television transmission signal*.

sound-programme Signals within a range of frequencies which human beings can hear (i.e. from 20 Hz to 20 kHz), and representing music or speech prepared for transmission as a radio programme, or as the sound component of a television programme.

149

sound-programme circuit A one-way *audio circuit* for the transmission of a *sound-programme* or the sound component of a television programme.

sound signal An *electrical signal* corresponding to acoustic signals in the range audible to humans; i.e. from 20 Hz to 20 kHz.

sound waves Waves transmitted through air (or any other gas, solid or liquid) as a result of vibration through the medium in which it is travelling. In communication systems, sound waves are generally considered to be within the range of frequencies which human beings can perceive; i.e. from 20 Hertz to 20 kiloHertz (20 Hz to 20 kHz).

source address field A *field* of *information* in a *frame* which identifies the *station* sending the frame to the *network*.

space In *alphabetic telegraphy*, one of two possible line conditions occurring in a *signal*, the other condition being known as a *mark*. All *characters* in telegraphy are made up of combinations of spaces and marks; i.e. patterns of *bits* represented by *binary digits* 0 and 1. In a transmission sequence, the first bit of each character is always a space which acts as a *start pulse*. The final bit is always a mark (a *stop pulse*), and the line condition remains as a mark until another character is transmitted.

space-division systems This term refers to systems in which *multiplexing* is achieved by *frequency modulation* (*FM*) rather than by *time division multiplexing* (*TDM*). Thus, terms such as space-division interface arise, to distinguish from a TDM *interface*.

space switched systems Used to refer to *switching* systems in which *frequency division multiplexing* is used, rather than *time division multiplexing*. ⇨ *switching equipment*.

space switching In *transmission systems* using *frequency division multiplexing*, the process of *switching* to make physical connections through an *exchange* is often referred to as space switching. This distinguishes from *time switching* which arises in systems using *time division multiplexing* (*TDM*). In both cases, a physical switching operation is still necessary, but furthermore with TDM, a *synchronization of time slots* is also necessary.

space-time-space network Relates to a *switching* process sometimes used in *digital exchanges* handling *time division multiplexed* traffic, in which incoming *signals* first pass through a *space switch*, then a *time switch*, and then a space switch. This method can be compared with the *time-space-time network* which is perhaps more commonly used. ⇨ *time shifting*.

space waves Radio broadcast *signals* used in short-range communication by means of *VHF* and *UHF* signals, in which the waves are propagated as a combination of direct and ground-reflected waves.

SPC Abbreviation of *stored program control*.

spectral colours Colours generated directly from the narrow *bands* representing the colour stages of the *visible colour spectrum*, rather than colours arising as combinations of these stages which are known as *non-spectral colours*.

spectrum envelope When a *signal* is *modulated* on to a *carrier wave*, a complex pattern of signals is generated; this would include the carrier wave plus *sideband* signals which are displaced in frequency on either side of the carrier and known as *upper sideband* and *lower sideband*. The total pattern of signals is known as the spectrum envelope. There are different methods for applying the basic modulation techniques resulting in different forms of the spectrum envelope. ⇨ *modulation*, *single sideband modulation* and *vestigial sideband modulation*.

speech channel A *communications channel*

designed to occupy a very *narrow bandwidth* equivalent to the range of sounds necessary for humans to speak to one another, i.e. an *audio frequency* bandwidth of 3 or 4 kHz.

speech circuit A *circuit* designed to carry *electrical signals* as a representation of human speech and offering a *bandwidth* of from 300 Hz to 3300 Hz.

speech communication Relating to a system designed to permit communication of human speech; i.e. to switch and transmit *electrical signals* up to 4 kHz in *bandwidth*. ⇨ *speech traffic*.

speech path A *transmission channel*, up to 4 kHz in bandwidth, used in a *telephone network* intended to carry *voice analog signals* representing human speech.

speech signal An *analog electrical signal* representing the sound of human speech. Same as *voice signal*.

speech traffic Relating to *messages* being transmitted to represent human speech in a *telephone system*, where each *speech channel* consists of *analog electrical signals* in a *bandwidth* of from 300 to 3300 Hz. Note, however, such signals can be converted to digital signals using *pulse amplitude modulation*, and therefore speech traffic can be carried by *time division multiplexing*.

spill-over traffic In a *switched system*, some traffic cannot be handled by the *basic route* during a *peak load*, and therefore is transmitted via an alternative *transmission path*. It is known as spill-over traffic.

split screen transmission A technique for entering *data* over a *network* in which an operator completes entries on to a form displayed on a *visual display* unit; but when the details are sent, only the variable data on the form is transmitted to the *line*. Also known as *partial transmission*.

SSB ⇨ *single sideband modulation*.

SSU Abbreviation of *subsequent signal unit*.

stability A quality desired in a *network* in which a high proportion of the *offered traffic* shall be delivered within the standards set, and that this proportion be maintained at all levels of loading planned for the network.

standard code sets A set of internationally agreed *codes* and their approved significance within an approved *data communication code*. Sometimes these code sets can be extended with non-standard codes used in local or national situations. The standard code is then a subset used for international purposes.

standing charge A charge made to a *user* for being a registered *subscriber* to a communications service. The charge may be levied as an annual fee, which may or may not include rental of a *terminal* and the allocation of reference numbers to allow access to facilities on the system. Contrast with *usage charge*.

star network A *network* in which the *users* in the system are connected by means of communication *links* radiating out, like spokes of a wheel, from a central hub or controlling *node* which handles all communication between users. ⇨ *network topologies*.

start bit A *binary digit* at the beginning of a series of *bits* representing a *character*. ⇨ *start code* and *start-stop signals*.

start code
1. A *bit* at the beginning of a series of bits representing a single *character*. The start code signifies the beginning of the character to maintain *synchronization* between the *transmitting* and *receiving terminals*.
2. Any bit sequence or character sequence used to signify the beginning of a unit of *information* in a transmission sequence. ⇨ *stop code*.

151

start of heading (SOH) An international *transmission control code*. This identifies a succeeding group of *characters* as a *header*; e.g. containing *routing* or other *control information* relevant to an ensuing *text*.

start pulse A pulse at the beginning of a *character* in an *asynchronous* transmission system and used to establish *synchronization* between *transmitting* and *receiving terminals*. ⇨ *start code* and *start-stop signals*.

start-stop signals Pulses used to indicate the beginning and end of a group of pulses which represent *information*. The pulses are used to synchronize the *receiving terminal* to the incoming stream of pulses representing information. For example, a standard *teleprinter* reacts to groups of seven pulses, five of which are coded to represent a single character of information, and two are the start and stop pulses. Each stop pulse is identified because it is of longer duration than the others. ⇨ *asynchronous operation*.

start-stop transmission A mode of transmission in which *characters* of *information* on the *line* are delineated by *start-stop signals* which synchronize the *receiving terminal* to the incoming stream of pulses representing characters. ⇨ *asynchronous operation*.

state One of the conditions which can exist during the establishment, maintenance or clearance of a *call*; e.g. *called terminal alerted state*.

station A location connected to a *network* and having a unique *address* to enable signals to be routed to it from other stations. A station may be able both to send and to receive *information*, and could be implemented as a *computer* or a *terminal*.

statistical multiplexor A development of the *TDM multiplexor* in which increased efficiency in the use of *data circuits* is achieved by giving increased priority to certain *channels* in the *transmission path* according to the expected transmission load. ⇨ *multiplexor* and *time division multiplexing*.

statmux Abbreviation of *statistical multiplexor*.

STD Abbreviation of *subscriber trunk dialling*.

stop bit A *binary digit* occurring at the end of a *character* and used to synchronize *transmitting* and *receiving terminals*. ⇨ *start-stop signals*.

stop code
1. A *bit* or series of bits at the end of a group of bits representing a single *character*. The stop code signifies the end of character to maintain *synchronization* between the *transmitting* and *receiving terminals*.
2. Any bit sequence or character sequence used to signify the end of a unit of *information* in a transmission sequence. ⇨ *start code*.

stop pulse A pulse occurring at the end of every *character* to establish synchronization between *transmitting* and *receiving terminals*. ⇨ *start-stop signals*.

store and forward systems Refers to a communication system in which *messages* may be transmitted, even though a complete *transmission path* to the *receiving terminal* is not available. The system will entail one or more *computer*-controlled *exchanges* (or *nodes*) which are able to store messages and release them for transmission to the receiving terminal when a transmission path is available.

stored program control (SPC) This concept refers to the application of *data processing* and *computer* control techniques in the operation of an *exchange*. The techniques can be applied to any service, e.g. *telex*, *telephone* or *integrated digital exchange*.

SPC provides the means of using a computer program to give *centralized control* over the operational, administrative and maintenance functions. It interprets *signals* received from *subscribers* and other exchanges, arranges the set-up and clearing procedures and paths through the switching unit, and determines what *messages* are sent to the various subscribers or other exchanges connected to the particular exchange.

Strowger Refers to Almon Strowger who developed a mechanical relay in 1891, the principles of which are still used in many countries as the basis of electromechanical *switching* operations in *telephone exchanges.*

Strowger selectors These are electromechanical devices used in early *circuit switched telephone exchanges* to make connections between *incoming lines* and *outgoing lines.* Each *selector* in such a system is activated by a *solenoid* or a motor. The system was first developed by Almon Strowger in 1891, but has seen many refinements. The benefits of the system are the reliability and simplicity of the selector devices which can be activated directly by *signals* originated from the telephone dialling mechanism. In recent times, Strowger exchanges have been replaced by *electronic exchanges*, which are more efficient when used in conjunction with digital forms of transmission.

S-T-S network ⇨ *space-time-space network.*

STX ⇨ *start of text.*

SUB A special *character* used in a *data communications code* to present a character which is to be substituted for an erroneous or invalid character.

sub-address A suffix added to a *network user address* to identify a particular *terminal*, *user*, or class of use available within a communications complex.

submarine cable A cable submerged beneath the ocean to provide communica-

tion between land-masses separated by sea. This is one of the earliest forms of *transmission link* between the major continents of the world, and such cables are still being laid down, despite the rapid growth of *satellite* communication. Submarine transmission links usually consist of *co-axial cables* with *repeaters* sealed into the construction of the cable at regular intervals. *Multiplexing* techniques are used so that a single *channel* can be used for thousands of simultaneous *calls.*

subscriber Any person or organization registered as a *user* of a *telecommunications* service and who is entitled to send and receive *information* over a *network.* The subscriber is usually identified by an *address* which is typically a *number* providing a unique identifier for a particular *terminal*, or a password related to a named individual or organization.

subscriber instrument A *terminal* registered for use by an individual or an organization over a *public network*; e.g. a *telephone.*

subscriber switching subsystem Provides for the concentration of *traffic* from lightly utilized *subscribers' lines* at a *local exchange* on to the heavily used common *circuits.*

subscriber trunk dialling A service provided to *subscribers* in a *telephone network* in which they are able to call distant subscribers over *trunk circuits* without the intervention of a human operator.

subscriber's line The *transmission link* between a subscriber's *terminal* and the *local exchange* with which it is permanently associated. Also known as *local line* and, in the USA, as *customer's loop.*

subsequent signal unit (SSU) In some systems, groups of *signals* may be transmitted as a unit known as a *multi-block.* Any signal, other than the initial signal of the *block*, is referred to as a subsequent signal unit.

superhigh frequency (SHF) A range of *electromagnetic waves* in the range of from 3 to 30 GHz, and having *wavelengths* from 1 to 10 centimetres.

supergroup In relation to *frequency division multiplexing*, an assembly of equipment providing a means of transmission for 60 separate *voice channels* along a single *coaxial tube*.

supervisory control language A type of *programming language* used by system managers responsible for customizing the facilities provided in a communications *network* based upon a *stored program control* computer.

surface waves *Radio waves* which are guided over the earth's surface and which usually occur in the *low frequency* and *medium frequency* bands of the *radio frequency spectrum*.

switch hook The contacts which are activated when a telephone *handset* is raised or set down. Sometimes called a *cradle switch*.

switched circuits *Circuits* which are formed by activating switches within an *exchange* to connect one set of *lines* with another. The implication is that a *switching* operation is performed to make a particular *call* possible, and that the connection is automatically maintained until either party *clears* (or rings off). Contrast with *packet switching* and *message switching*.

switched connection A *transmission path* established by an *exchange* and entailing the physical connection of *circuits* in the exchange by means of electro-mechanical switches, or entailing the *synchronization* of timing between an incoming and an outgoing *digital channel* to complete an *access path* between a *calling terminal* and a *called terminal*.

switched system ⇨ *switched telecommunication system*.

switched telecommunication system In a switched telecommunication system, there are three basic elements: *terminals*, *transmission links* and at least one *exchange*.

The objective of a switched system is to enable a *user* to communicate with any other user at any time.

In the *telephone system*, a user is able to talk to another by dialling a connection from his own terminal, via a *local line*, to his *local exchange*, and thence by transmission links (possibly including other exchanges) to the user required. Once a *transmission channel* is established between the two users, it remains available until the *call* is finished. This is an example of a *circuit switched system*.

There are, however, switching systems in which transmission links are used but not allocated exclusively for the duration of a particular call. Such systems are more often used in the transmission of *data* and may be *packet switched* or *message switched* systems.

In a packet switched system, data is transmitted in *blocks* which contain identifying information, allowing the *packet* to be associated with a particular call to a particular user. As they pass over the transmission links, the packets forming the elements of a *message* may be interspersed with packets related to entirely different calls, destined for other users. No continuously available *channel* exists between two terminals in such a system, yet such systems, working at high speeds, can conduct communications in *conversational mode*, as well as carry messages from a source to a destination. (⇨ *packet switching networks*.)

In a message switched system, a user can transmit a message to the network, where it may be stored and delivered to the intended recipient at a later time.

Telegraph systems were originally conceived as circuit switched systems, but were later developed as message switched systems also. In a telegraph system, a transmission channel can be established between two *teletypewriters*, providing a physical connection for the duration of the call to allow communication in conversational mode or message transmission. A permanent record (e.g. in *punched paper tape* or *hard copy*) is created during the call at each terminal.

Teletypewriters are slow-speed devices and are used over *networks* at speeds ranging from 50 to 150 *bauds*. When used over message switched networks, the time taken to deliver each message renders the system unsuitable for conversational mode. It is common for *message queues* to occur and, on occasions, messages can take several hours to reach their destination. Messages are usually stored in *computers* until a suitable transmission link is available to complete transmission to the destination. Sometimes such systems are called *store and forward systems*.

Message switched systems can, therefore, be used to attempt to optimize the cost of the telecommunications activity by delivering messages by an available *route*, taking advantage of differential *tariffs* which may apply in different time zones and through different *carriers*.

The distinction between a message switched system and a circuit switched system is determined largely by the design of the exchange.

An exchange which establishes a physical transmission channel between two terminals for the duration of a call is known as a *circuit switched exchange* and is necessary for person-to-person communication.

An exchange which stores messages until it can establish a transmission channel is a *message switched exchange*, and is generally used for transmission of messages or small units of data. ⇨ *switching equipment*.

switched telephone exchange A *node* in a *telephone network* in which *circuits* are connected automatically in response to *signals* generated by *users* dialling *addresses* upon a telephone instrument.

switched virtual circuit Same as *virtual circuit*.

switching equipment For many years the technology used in *telephone exchanges* to switch connection between *circuits* was based on electro-mechanical devices, made up of rotary switches and called *selectors*. The concept of the automatic telephone exchange, using such principles, was invented in 1889 by an American undertaker, Almon B. Strowger. Although there have been advances in the technique, the basic principles have remained in use until the present day and the name Strowger is still used to describe such systems.

With the Strowger principle, it takes a few seconds to make an automatic connection but, with the increase in the number of telephones and the spread of *international trunk dialling* with long numbers, this system now appears to be slow and inefficient to *subscribers*.

An improved electro-mechanical system, known as *crossbar exchange*, was introduced in the late 1920s. In a crossbar system, a complete matrix of wires exists in the exchange, so that to make a connection, the vertical wires from an *incoming call* are linked to the horizontal wires of a desired outgoing circuit at the point (known as a *crosspoint*) where the wires cross in the exchange. This is an oversimplified description, but it introduces an important feature in exchange development: the concept of a *common control* which recognizes requests and makes connections.

Crossbar exchanges have made rapid progress in many industrialized countries and, with the introduction of *electronic exchanges*, the principles of crossbar systems have been adapted most successfully to provide the design basis for the early electronic switching systems. These electronic exchanges have the advantage of relative cheapness and are more easily maintained than electro-mechanical systems, as well as having the ability to record activities automatically; e.g. for providing *billing information* or *traffic statistics*.

Many electronic exchanges have *digital computers* controlling them, and the common control features of the crossbar exchange are now provided by computer programs in what is known as *stored program control (SPC)* exchanges.

SPC exchanges are very flexible; for example, simple instructions to the computer program can be used to make changes which previously had to be made as physical connections by an engineer.

There is now a rapid change taking place in exchange and transmission technology, and this is concerned with the introduction of *time division systems* which use transmission techniques based upon *time division multiplexing (TDM)*. Such systems must use *digital transmission*, not *analog* transmission, and TDM was first applied to *data networks*. In time division systems, a number of *calls* are able to share the transmission highway; in effect, the use of the *highway* is available as a series of *time slots*.

A time slot might be 8 *bits* in length and, typically, 30 active channels might share a 2 Mbit highway.

Now, time division techniques have been extended to *voice* networks in which each human *conversation* is converted to a string of *binary digits* for *transmission* and reconverted to audio *analog signals* for the *subscriber*.

The crosspoints in such a TDM system are referred to as *time shared digital crosspoints*; to connect two such channels in an exchange requires two processes: *space switching* (i.e. making the connection from one channel to the other), and *time switching* (i.e. *synchronizing* the time slots occupied in the two channels).

It is common to refer to TDM systems as *time switched systems* and to non-TDM systems as *space switched systems*. Time switching techniques enable several conversations or messages to share the same line; they can likewise efficiently time-share the crosspoints in an exchange.

TDM systems are expected gradually to replace the so-called space switched systems; but this will take time, due to the heavy investments made by *telecommunications* authorities in the earlier systems. There are, however, great efficiencies in the utilization of lines to be obtained with time switching, and a far greater potential to extend a whole range of services, digital, voice, and *graphics* to the individual office or home.

Telecommunications networks will, in future, combine speech, *data* and other services; this concept is referred to as *integrated digital transmission and switching*. The following are included in the benefits to be derived from it: fast set-up for *multilink calls*, common *signalling systems* with high capacity, transmission loss independent of the length of circuit, low *distortion* and *noise*, high reliability, and an extensive selection of services to the subscribers.

All the topics discussed above are examples of *circuit switching*. ⇨ the article entitled *switched telecommunication system*; that article also describes *message switching*, which operates on different principles.

SYN ⇨ *synchronous idle.*

sync bytes Units of *information* which precede each transmission in a *synchronous* system, to notify the *receiving station* that a *signal* is about to be transmitted. The *receiver clock* in the *receiving terminal* uses these pulses to establish *synchronization* with the *transmitter clock*.

sync pulse ⇨ *synchronization pulse.*

sync pulse separator A device in a *television receiver* which is used to extract *synchronization* information representing *field scan* and *line scan* instructions, so that a scanning operation can be generated to produce an accurate picture.

synchronization The process by which *data terminals* connected to a *network* can communicate with one another. The methods vary from one class of system to another, but the principles are discussed under the following headings: *bit synchronization*, *character synchronization* and *message synchronization*.

synchronization pulses In a television system, these are pulses generated during the camera scanning process, transmitted as part of the *composite video signal*, and used to maintain *synchronization* between the camera and the *TV receiver*. In practice, these pulses enable the *scanning beam* in the receiver to be related to the equivalent scanning points traced by the scanning beam in the camera. ⇨ *video signal*.

synchronization signal unit (SYU) A unit of *signal* information, which contains *information* to synchronize a number of other *signal units* transmitted as a *multi-block* on a *signalling* channel.

synchronized network A *network* in which the timing of all operations is governed by one *master clock*. Also known as a *despotic network*; contrast with *democratic network*.

synchronizing pulse ⇨ *synchronization pulse*.

synchronous data link control (SDLC) A *line control* protocol developed by IBM which is a subset of the *high level data link control* (q.v.) developed by the International Standards Organization. It performs transmission *error checking* functions and automatically requests retransmission of error *frames*.

synchronous data transmission A system used for high-speed *digital transmission*, in which *synchronization* is maintained by *clocks* in the *transmitting* and *receiving terminals*, and *binary signals* are transmitted without gaps. This continues for as long as necessary and, if there are gaps in the *data*, the transmitting terminal must insert idle *bits*. Synchronization is established at the beginning of each transmission by special *SYNC* characters which are used to align the clocks at each *station*. Contrast with *asynchronous operation*.

synchronous detection A method of *detection* used in *demodulation* where the method of transmission involves suppression of the *carrier* and *upper sideband* signals. It involves the generation in the

demodulator of a local *carrier signal*. ⇨ *modulation*.

synchronous digital system Any communication system in which communication is in *bits* and in which *synchronization* between *transmitting* and *receiving stations* is achieved by arranging for their *clocks* to be synchronized following the transmission of *synchronization pulses* at the beginning of each *message signal*. Contrast with *asynchronous systems*, in which timing is established by *start* and *stop pulses* associated with each unit of *data* (e.g. each *character*).

synchronous idle An international *transmission control code*, transmitted to establish or maintain *synchronization* with a *receiving terminal*.

synchronous mode Relating to communication using *synchronous data transmission*.

synchronous operation A method of *data transmission* in which *characters* are transferred as a contiguous sequence of bits and there are no gaps or *start* and *stop bits* between each character. The receiver establishes *synchronization* with the first bit of any character sequence and maintains synchronization with a clocking mechanism which maintains a constant frequency with the periodic time of the incoming *bit stream*. Contrast with *asynchronous operation*.

system boundary The physical limits of a particular system, excluding the *users* at the *transmitting* and *receiving stations*, but including the *terminals* and the intervening *transmission channel*.

system control signal unit (SCU) A *signal* transmitted on a signalling *channel*, but having the purpose of changing the operation of the *signalling system*; e.g. to switch the signalling *traffic* to a designated *link*.

systems network architecture (SNA) A

SYU

protocol and implementation for a *data network* developed by IBM in the 1970s for providing end-to-end management of *access paths*, and representing one of the first realizations to allow *terminals* in a *tree network* to share a *line* to separate *applications* in the same *host processor*. Later features provided *computer networking* which allowed any terminal to gain an access path to any host processor in the network by the interlinking of tree networks.

SYU Abbreviation of *synchronization signal unit*.

T

tandem exchange A particular method for linking a number of *local exchanges* in a *telephone network* so that *traffic* may be carried between these local exchanges without having to be *routed* via *trunk circuits* and *primary trunk exchanges*. This practice is often adopted in city areas in which there is high *traffic volume* and high *subscriber* density. Also known as *tandem office*.

tandem office ⇨ *tandem exchange*.

target probability of service The targets set by the designers of a system to provide service to the *users*. For example, in a *circuit switched system*, the target probability of *call blocking* may be set as 0·01.

target user population The group of people or organizations for whom a particular service is designed and promoted; e.g. the residential community in the United Kingdom, or commodity brokers in nominated European cities.

tariff The charge made to *users* by a supplier of communication services for use of a communications *network*. This normally entails the users being registered and given a *customer reference number* for *billing* purposes. The identification and recovery of charges is often a significant cost in the development and operation of a service. ⇨ *usage charge, standing charge* and *connection charge*.

TC Abbreviation of *transmission control code*.

TD Abbreviation of *time division*.

TDM Abbreviation of *time division multiplexing*.

TDM multiplexor A *multiplexor* which uses the principles of *time division multi-plexing* to combine a number of *channels* on to the same physical *transmission path*.

telecommunication Communication between two people using equipment to overcome the effects caused by distance and/or physical barriers between them. The communication need not be two way, and the equipment need not necessarily be electronic.

teleconferencing A system of communication designed to enable a group of users to be linked up using telecommunication facilities in such a way that they can conduct a conference or meeting although they are all situated in different geographical locations and, perhaps, different time zones.

This system has primary applications in the business field, for example, to enable a group of executives to discuss an agenda and make decisions without travelling to meet at a specific location. Another example occurs in distance teaching in which a lecturer is able to address a group of students who cannot be effectively brought to a single location.

A conference can be established using dial-up telephone facilities, but *computer conferencing, picture phones*, and *remote electronic blackboards* may be used.

telegraph codes Telegraph codes grew from the need to transmit plain language *text* and numerals, and telegraph codes were designed to represent the letters of the alphabet and numerals. Thus, the early *codes* contained just 36 *characters* or codes. These characters are represented as groups of electrical pulses during transmission, and *messages* can be prepared and stored as holes punched in *paper tape*. Each row of holes in the paper tape represents a single character, the *pulse codes* being represented by the presence or absence of holes in each row.

In order to incorporate punctuation marks and other special codes, the *character set* was expanded beyond the basic 36 characters, and the present international standard set (known as *International Alphabet No. 2*) contains 58 different characters.

It is a *5-bit code* (each character is represented by a combination of 5 pulses). Strictly speaking, such a code only provides 25 or 32 unique characters, but *shift characters* are used to extend the capacity of the code; i.e. a *figures shift* code indicates that all following characters are to be treated as figures or other *secondary characters*, while the *letters shift* signifies that following codes are letters.

In telegraph communications, each character is framed by a *start pulse* and a *stop pulse*. Thus, 7 bits are actually transmitted to line for each character.

With the development of computer-based systems, telegraph codes have expanded to meet further requirements. ⇨, for example, *data communication codes* and *International Alphabet No. 5*.

telegraph exchange
1. Refers to the widely used international service, which is based on the transmission and reception of printed *messages* over a *telegraph network* using *teletypewriters* as subscriber *terminals*. Also known as *telex*, a contraction of *te*legraph *ex*change.
2. May also be used specifically to refer to an *exchange* (or *node*) which performs the *switching* operations in such a *network*.

telegraph network A *network* designed for the transmission of telegraph *traffic*, and comprising telegraph *terminals* (e.g. *teletypewriters*), *transmission links* and automatic *exchanges*.

telegraph service A service provided by a *telecommunications* authority for carrying one or more kinds of telegraph *traffic*. Such services may use different techniques, including *alphabetic telegraphy*, *document*

facsimile telegraphy, *photographic facsimile telegraphy*. These services may also be known to the public by specific names which characterize the nature of collection and delivery, such as *telex*, telegram and radio telegram services.

telegraph system A system in which *users* have *terminals*, known as *teletypewriters* (often the trade name *teleprinter* is used), to transmit and receive *messages*. Teletypewriters are fitted with *keyboards* and a *printer mechanism*, and sometimes they can create and read *punched paper tape*.

The *message signal* output from a teletypewriter is a *digital signal* which is transmitted along a *line* to operate another teletypewriter remotely. The speed of transmission is usually in the range of from 45 to 200 *bauds*.

A range of *alphanumeric* characters can be transmitted and, on standard equipment, each *character* is transmitted as 5 pulses to represent the character plus a *start* and *stop pulse*.

Teletypewriters operate over *networks* controlled by a *telegraph exchange* and often the term *telex* is used to refer to systems of this type. More strictly known as *alphabetic telegraphy*.

telegraphy A means of two-way communication for the transmission of *messages* between two locations, including: (a) the transmission of *alphanumeric* information using *teletypewriters*, which is a form of *alphabetic telegraphy*; (b) the transmission of *electrical signals* representing the images scanned from documents, and which is known as *document facsimile telegraphy;* (c) the faithful reproduction of continuous tonal densities from documents to a quality allowing good photographic images to be transferred – known as *photograph facsimile telegraphy*.

telegraphy, alphabetic ⇨ *alphabetic telegraphy*.

telemetry The measurement of events at a distance (sometimes great distance as in aerospace projects), in which *transducers* are used to measure physical activities and to convert these to *digital signals* which reflect the measurement of phenomena. The digital signals are transmitted to a *data collection* centre for processing. Sometimes the results of such processing are immediately used to generate signals which are transmitted back to the distant location to control the process being monitored.

telephone
1. Refers to the *public switched telephone network (PSTN)* which is a system for transmitting *voice signals* in the *audio frequency* range over a *network* servicing a very large *user* population.
2. Also refers to the telephone *terminal* or telephone *handset* used as the primary *interface* between human users and the system.

telephone channel A *transmission path* designed for the transmission of *signals* representing human speech and having a *bandwidth* of about 3 kHz.

telephone exchange A set of equipment designed to connect *subscriber instruments* to one another over *transmission links*, to enable telephone *conversations* to take place. A feature of a *switched telecommunication system*.

telephone system A system for two-way communication in which a *user* is able to speak via a telephone *terminal (subscriber's instrument)* to another *user*, by dialling the *routing* required through an *exchange*. The exchange connects the caller to the required user by selecting appropriate *transmission links*.

telephony The technology associated with the transmission of *message signals* in the *audio frequency waveband* between *users* at distant locations, including the techniques associated with the transmission of *voice signals* by *digital transmission* methods.

teleprinter A registered trade name, but now used by many people to refer to the *teletypewriter*.

teleprocessing A term used to define the *application* of *data communication* techniques to *data processing* to create *on-line* or *real-time processing systems*. Generally speaking, a teleprocessing application operates in an environment in which multiple *users* are able to share a common *database* and the *hardware* resources and *software* facilities provided in a *main frame* or *minicomputer* system, without any user being aware of the activity of other users. The term has also been used to define a system in which programs are downloaded from a main frame to a *terminal*, to be executed by a *microprocessor* in the terminal under local control of a user.

teleprocessing monitor (TPM) A *software* system specially designed to ease the development of *on-line* or *real-time* computer *applications*. Generally speaking, the TPM is the *interface* between *user applications* and the general-purpose *operating system* of a *main frame* or *minicomputer*.

The TPM is designed to create an environment in which many of the problems peculiar to on-line computer applications can be easily resolved. These problems include: (a) multiple users sharing the same *application programs*; (b) screen formats and editing and validation procedures being easily set up and altered without detailed changes of *application code*; (c) the ability to amend easily the conventions associated with *terminals*, thus leading to flexibility in the number and type of terminals used; (d) *lockout facilities* to ensure the integrity of *data*, where several users may be amending *files* or reserving objects in the same *database*; (e) the establishment of *recovery procedures* which preserve the integrity of the database, or minimize the work and time involved in recreating a working system after a system failure.

A TPM may be supplied by a *computer* manufacturer with its *hardware* system, or be supplied by specialist companies. It may provide many more facilities than those described above, but its principal objective is to make it easy for those who wish to develop complex *data processing* systems

161

providing *data communication* facilities for multiple users.

Teletel The name given by the *PTT* in France to its *public videotex system* planned for introduction in France in the early 1980s. ⇨ *videotex system.*

teletext A cheap system of mass communication in which the *user* is able to receive *text* and *graphics* using a specially adapted *television receiver* to respond to *signals* broadcast in the blank intervals of a *television transmission signal.* Compare with *videotex system.*

teletext decoder A micro-electronic device fitted to a *television receiver* to enable it to detect and display *signals* representing *text* and *graphics* which are transmitted in the blank intervals of a *television transmission signal.*

teletext editing terminal A *terminal* designed for creating *graphics* and *text* in accordance with the standards for a *teletext* service. Used by an *information provider* to create or update a *database* which is to be *broadcast* to teletext *subscribers.*

teletypewriter A device used to transmit and receive *information* over a *telegraph network* and consisting of a printer and a *keyboard*, with a transmitter and receiver to *interface* with a *line* and a *telegraph exchange.* The keyboard is similar to a typewriter but has special function keys to control its specific use in a *telegraph system.* A teletypewriter is able to print a full range of alphabetic, numeric, and punctuation *characters*, plus special symbols. When the keys are depressed, a series of direct current electrical pulses are generated to represent the selected characters. These *signals* are used to activate the printer unit and to transmit to the line and can be thought of as *digital signals*, where each pulse represents a *binary digit.*

The printer can also be activated by signals received remotely from the keyboard of another *subscriber.* Thus, a tele-

typewriter can be used to conduct a *dialogue* with a distant subscriber.

The speed of operation is usually fixed at 50 *bauds* (i.e. 50 *binary digits*) per second, which is less than 7 characters per second since each character consists of a *start pulse*, five pulses to represent the character itself, and a *stop pulse* of longer than standard duration. Some teletypewriters have been made to operate at higher speeds, up to 300 bauds per second.

Standard international codes are in use today and have been promoted by the *International Telecommunication Union* (*ITU*) to achieve compatibility between the different countries of the world.

television channel A path of communication for *radio signals* carrying television transmission and appropriate *audio signals.* The channel may occupy a particular *bandwidth*, either as a radio broadcast in the atmosphere or as a cable broadcast. A 625-line television channel usually has a bandwidth of approximately 8 MHz in the *UHF* range, *vision* and *sound* information being transmitted on separate carrier waves which are separated within the channel width by about 6 MHz. ⇨ *television transmission signal.*

television circuit A one-way *circuit* for the transmission of the video component of a television programme. The sound component is usually delivered on one or more separate *sound-programme circuits.*

television receiver A *terminal* designed to receive pictures and *sound*, transmitted as a *radio signal.* The receiver incorporates a radio frequency *filter* and *amplifier*, a *demodulator*, *video amplifier*, *audio amplifier*, *cathode ray tube* and *loudspeaker.*

Normally, the *signal* is received as a radio frequency *broadcast* coming from a *television transmitter*, but signals can be received along *co-axial cables.*

The filter and the amplifier are *tuned circuits*, used to select the appropriate *transmission channel* required by the *user.*

The demodulator selects the *message signal* and separates it from the *rf carrier wave.* This signal, along with appropriate *synchronizing pulses*, is fed to the video and audio amplifiers, where the message signal is amplified to drive the cathode ray tube and the loudspeaker system.

TV receivers within a country are built to common standards, but there are 3 major colour systems in use today: *NTSC* – USA, Canada, Japan; *SECAM* – France, Soviet Union; *PAL* – United Kingdom and West Germany. Variations of these basic systems have been implemented in other countries.

Early TV receivers could only display pictures in black and white, and grey scales. These are known as *monochrome receivers*. *Colour television receivers* have been progressively introduced since the mid 1950s.

television transmitter A device which can accept *video signals, audio signals*, plus *synchronization pulses*, and modulate these on to a *carrier wave* for transmission as a television *radio signal* by an *antenna*. ⇨ *television transmission signal*.

television transmission signal A *radio signal* (usually transmitted in the *VHF* and *UHF* spectrum) which includes video and audio information transported on separate *carrier waves* which are combined for the purpose of transmission. The *signal* would normally be *broadcast* in the atmosphere or loaded into a *TV cable system* for transmission to *TV receivers*.

The precise characteristics of a transmitted signal may vary from one country to another, but usually television broadcasts occur in *bands*, e.g. 470–610 MHz. In the British 625-line system, these bands are divided into 8 MHz sections called *television channels* and a broadcast occupies one such channel. There is a *guard band* at the upper end of each channel and the *sound carrier* is transmitted close to this band. The sound carrier is separated by 6 MHz from the *vision carrier*. The *sound bandwidth* is from 20 Hz to 15 kHz and the *video bandwidth* is 5·5 MHz. The *video signal* and *audio signal* are *modulated* on to separate carrier waves which are combined

and normally broadcast from the same *antenna*. Usually *amplitude modulation* (*AM*) is used for the video component, and the audio component may be *AM* or *FM*.

In order to economize in bandwidth and transmission power, the *lower sideband* and the *upper sideband* are not both transmitted. Instead, most of the lower sideband and the vision carrier are suppressed. A portion of the lower sideband (known as the *vestigial sideband*) is transmitted, along with the complete upper sideband. *Vestigial sideband modulation* is chosen to simplify the circuitry required in TV receivers.

telex The name often given to a service based on the use of *teletypewriters* to send and receive *information* over a *switched network*. The word itself is an abbreviation of *telegraph exchange*. This system is used extensively throughout the world and there are many national and international telex exchanges. ⇨ *telegraph system*.

telex traffic *Message signals* derived from *teletypewriters* and transmitted in accordance with the international conventions of *alphabetic telegraphy*; e.g. using *International Alphabet No. 2*.

Telidon The name given by communications authorities in Canada to the planned *public videotex system* and, more specifically, to describe the Canadian standard of *alphageometric coding* used to provide a *high resolution display*. ⇨ *videotex system*.

terminal A device used in a *telecommunication* system for sending or receiving *message signals*. A *microphone* is an example of a *transmitting terminal* and television is an example of a *receiving terminal*. Some terminals (e.g. the *telephone*) can be used for sending or receiving message signals. A *teletypewriter* is an example of a terminal which, given appropriate *transmission channels*, can carry out simultaneous two-way transmission.

There are a great many terminal types to

be considered, from the simple telephone terminal to the large general-purpose *computer*.

terminating traffic This term defines the use made of a particular *terminal* by *incoming calls*, and contrasts with *outgoing calls* referred to as *originating traffic*. ⇨ *traffic, instantaneous traffic* and *traffic volume*.

tertiary trunk exchange The third level of *exchange* in an *exchange hierarchy* and controlling major *routes* in the *trunk network*. ⇨ general article on *exchange hierarchy*.

text In *telecommunications*, this term is used to refer to the parts of a *signal* which carry *user* information, rather than any *control information* which is embedded in the *message signal* to provide *link control* or *routing* instructions.

theoretical final route A term used in *telephony* to define the *route* followed by a *call* through the *circuit groups* of the *backbone* of the *network*.

3-bit error An error occurring in a *data transmission* sequence in which three consecutive *bits* are incorrectly received. ⇨ *transmission errors*.

time division multiplexing (TDM) A form of *multiplexing* in which a number of separate *digital signals* are sampled to produce a set of interleaved pulses which are transmitted along a single *channel*. The individual digital signals are combined to a signal of much higher *bit rate* for transmission. The train of pulses is analysed at the other end of the channel by logic *circuits* which recreate the original *message signals*. The technique of multiplexing may differ for particular systems, but usually the digital signals making up the *characters* or pulses of each message are grouped as units of short duration (bits/bytes), which are transmitted along with *framing bits* providing terminal identification and refer-

ence information for the transmission and receiving equipment.

This technique is so called because the elements of different message signals occupy the same *transmission channel* but for different moments in time. Thus, the process of *switching* such circuits requires both a physical connection of channels (*space switching*) and a *synchronization* of *time slots* (*time switching*).

This technique can be contrasted with *frequency division multiplexing* (*FDM*) which is separately described.

time division system A communication system which utilizes the techniques and principles associated with *time division multiplexing* (*TDM*).

time interval A brief interval in time during which a *signal element* (i.e. a *bit* or pulse) may occur and be recognized. In a *synchronous system*, each *character* may be composed of 8 bits; thus, 8 time-intervals are required to transmit a single character. In an *asynchronous system*, a *start bit* (one time interval) and a *stop bit* (two time intervals) are needed, in addition to the *information bits*. Thus, there are 11 time-intervals per character.

time shared digital crosspoint An element in a *digital exchange* which enables connections to be made to connect *incoming calls* with *outgoing lines*, but one which services a system based on *time division multiplexing* (*TDM*) techniques. Thus, a *switching* operation entails a physical connection of *channels* (*space switching*) and the *synchronization* of *time slots* (*time switching*). ⇨ *electronic digital crosspoint*.

time sharing Relating to any *computer* or communication system in which a device, such as a *computer* processor or a *transmission channel*, is used to control or support a number of separate processes. The priority is given to each process in turn for a few milliseconds, but the overall impression to a human observer is that all processes are operated simultaneously.

time sharing link A *communication channel* in which several *subscribers* share the same physical *line* by occupying the line momentarily in bursts of activity occupying a particular *time slot*. Each *user* is unaware of the others sharing the same resource. The techniques used are known as *time division multiplexing (TDM)*.

time shifting An operation which takes place in *time division multiplexing* systems to adjust the *time slot* of an incoming *channel* to the required time slots for an outgoing channel. If the time slots are already synchronized no shifting is needed, otherwise the usual technique is to delay the incoming channel, until the next time slot for the outgoing channel occurs. This is achieved by placing incoming *bits* into a *buffer* and releasing the bits through an electronic switch to an outgoing buffer at the appropriate time slot.

Usually, *digital exchanges* are constructed to combine *time switching* and *space switching*, the latter being achieved by shared *electronic crosspoints* which are activated for a few microseconds at the appropriate time for each time slot. ⇨ *time-space-time network*.

time slot A common concept in all *digital systems*, a brief period of time in which a particular device is able to obtain access to another to deliver or accept *digital signals*. For example, in *time division multiplexing*, each physical *communication channel* is *time shared* by many signals, each having access to the channel for a prescribed interval of a few microseconds on a cyclic basis.

time switch A device which controls the utilization of a resource by competing processes, by allocating a *time slot* of a few milliseconds or microseconds to each process.

time switched systems A switching system used to create *networks* for *digital traffic* (which may be derived from *voice* or *data sources*) and in which the techniques of *time division multiplexing (TDM)* are used.

time switching An operation which takes place with *time division multiplexed* systems to switch *signals* from one *time slot* to another. ⇨ *time shifting*, and contrast with *space switching*.

toll centre An *exchange* in a *telephone network* which provides the connections from a group of *local exchanges* to the *trunk network* of the overall *telephone system*. A *subscriber* thus gains access to the national network via his local exchange and the associated toll centre, and pays *toll charges* for *outgoing calls*.

Also known as *primary trunk exchange*, *primary centre* and, in the UK, as *group switching centre*. ⇨ *exchange hierarchy*.

toll charge An additional charge levied to a *subscriber* when he/she originates a *call* to an *address* outside the *local area* in which local call charges are applied.

toll circuit A term used to describe specifically the *circuit* between a *local exchange* and a *trunk exchange* in a *telephone system*. In Britain, the term *junction circuit* is often used to describe this type of circuit, but more generally the term *trunk* is used. It is, however, the circuit which provides the threshold from the local exchange to the full *trunk network*.

toll network ⇨ *trunk network*.

toll office ⇨ *trunk exchange*.

torn tape centre A centre in which a number of telegraph *circuits* terminate and in which *messages* are switched by human operators who remove *paper tapes* from receiving devices and transmit them over other circuits. Such centres have largely been replaced by automatic switching centres.

TPM ⇨ *teleprocessing monitor*.

traffic This term is used to define the utilization or capacity of a *telecommunications* system, and is described in terms of the number of *calls*. For example, the number of calls in progress at any one time is referred to as *instantaneous traffic*. The average value of instantaneous traffic over a given period of time is the *average traffic*.

Traffic is measured in units called *erlangs*; e.g. an instantaneous traffic of 57 calls would be described as 57 erlangs. ⇨ *instantaneous traffic, busy hour traffic, originating traffic, terminating traffic, average traffic, carried traffic* and *offered traffic*.

traffic analysis Information resulting from a formal observation of the performance of a *telecommunication* system, and intended to assist the system operator to plan improvements in the performance of the system and the service experienced by *users*.

traffic overload A situation of poor service which arises when the *traffic* offered to a part of a communication system exceeds, significantly, the capacity planned in the chosen design.

traffic pressure A measure of the demand for connections in a switched network, usually expressed as the number of *bids* made to secure a connection over a given *group* of *circuits* in a defined period; e.g. bids per circuit per hour.

traffic recorder An item of equipment used to measure the utilization of a particular part of a *circuit switched system*.

traffic statistics *Information* which has been collected and summarized to represent the pattern of usage of a communications system. Such information is used to plan for the efficient utilization of the *network* and to achieve a desired level of user service. ⇨ *traffic* and *traffic theory*.

traffic theory A theory to assist in the planning and development of a communications system in which the planner tries to optimize the equipment and facilities needed to support the requirements of *users*. A system which has too large a

capacity for a given population of users is wasteful of resources, but a system which does not have sufficient capacity will become congested and give a poor service.

Traffic theory usually requires planning engineers to consider the average behaviour of users of the system and relies upon the fact that, in a large enough population, the random behaviour of individuals tends to present a consistent pattern.

Traffic theory includes consideration of the following:

number of *calls* in a period (n)
the holding time of each call (h)
the duration of the period (T)
the rate at which new calls arise (a)

In practice, systems are usually planned to provide capacity to suit the *peak load* generated by the users.

Some equations used in traffic theory are given as follows:

The average traffic E for period T is: $\dfrac{nh}{T}$

or $E = ah$

The average rate for new calls (a) is given by: n/T. ⇨ *traffic volume, mean holding time*.

traffic unit A unit employed in the measurement of the use made by the *users* of a *telecommunication* system. Usually the *traffic* generated by users is measured in terms of the number of concurrent *calls* (⇨ *traffic theory*). However, for the purpose of *billing* users of the system, a system operator may also designate a unit of time as the basis of accounting; e.g. a *local call* may be charged in increments of charge per minute.

traffic volume The amount of *traffic* over a given period of time in a *telecommunication* system. If in T seconds n *calls* are made and if the duration of the calls are $h_1, h_2, h_3 \ldots h_n$ seconds, then the utilization of the system is:

$$\sum_{i=1}^{i=n} h_i$$

The *average traffic* (E) is the amount of

traffic in a given period of time and is expressed as:

$$E = \frac{\sum_{i=1}^{i=n} h_i}{T} \; erlangs$$

⇨ *mean holding time.*

transaction A specific *application* process in a *data network*, in which there is interaction between *transmitting* and *receiving stations* to perform a specific task; e.g. to record receipt of a payment and to credit a *subscriber's* account. ⇨ *session* and *datagram.*

transaction log A record of all the activities or *transactions* performed by a *computer*-controlled system. This record may be used to compile statistical information to help in optimizing the operation of the system, and also to recover from failures which might otherwise cause *message signals* to be lost.

transceiver A device which can transmit and receive *information.*

transceiver cable A cable used to connect a transmitting/receiving device to a physical *channel*. The cable must have a compatible *physical* and *electrical interface* to the *transmission channel.*

transducer The element in a *telecommunication* system which converts *signals* into a form required by *users*. For example, a *loudspeaker* converts *audio frequency* electrical signals into *sound waves*. A *microphone* is an example of a transducer which converts sound energy into *electrical signals.*

transient copy A *message* produced in a communication system, but not in a physical form; e.g. a message displayed on a *visual display*. Also known as *soft copy*, and contrast with *hard copy* or *permanent copy.*

transit exchange Any *exchange* in a *network* which acts for a particular *call* as an intermediate switching centre; i.e. not the

first exchange in a *routing* and not the *last exchange.*

transit network A *network* superimposed upon another for a special purpose. For example, in the UK it was once not possible to provide *STD* to all *subscribers*, due to inherent limitations in old exchange equipment which could not all be replaced at one time. A transit network was created to provide special routes to *group switching centres* capable of STD operations and thus overcome the dialling limitations.

translator A *control* device used during the set-up of a *call* to translate the *address* and determine the *routing* required.

transmission break A particularly significant form of error condition during which *line* signals are lost entirely during transmission of *message* traffic. Breaks of this nature are not uncommon in systems using older type of *switching equipment* where maintenance can be a problem. ⇨ *transmission errors.*

transmission channel The path between two *terminals* for communication and including relevant *links* and *exchange* equipment used to handle the *call*. ⇨ *data channel* and *data circuit.*

transmission control An element in a *node* of a *packet switching system*, which is responsible for interfacing *users* to the *network*. It assists in the management of establishment and termination of a *session* for an end user.

transmission control code Throughout the evolution of *data transmission* it has been common practice to send *message signals* framed by *control codes* which provide instruction to *terminals* as to the format and structure of message *texts*, and to provide *device control* information. The use of these *codes* has varied with advances in communications technology. Standards have been developed internationally and unique *bit patterns* to represent such codes are included in common *data communication codes* such as *ASCII, ISO 7,* and *EBCDIC*. Examples of the meaning of transmission control codes appear throughout this book. ⇨ *ACK, NAK, ENQ, SOM, STX, EOM, ETX, ETB, EOT, SYN* and *DLE*.

transmission errors Errors arising in a *message signal* and caused by *distortion* or disturbance on the *transmission channel*. All *telecommunication* systems are prone to errors, and it is necessary to design systems which automatically identify and recover from errors (⇨ *error checking*). The quality of a *circuit* is usually expressed in terms of its *bit error rate*, and this is a statement of the maximum average number of *error bits* to *correct bits* which can be expected when transmitting *digital signals* over the circuit. Some typical bit error rates are: (a) circuit at 1,200 *bauds* over the switched public *telephone network* – 1 bit in 10^3; (b) 2,400 bauds over a *private leased circuit* – 1 bit in 10^7.

Very often transmission errors are manifest as the simple reversal of a single pulse; e.g. a 1-bit is received instead of a 0-bit. This is known as a *single-bit error*. If a series of consecutive pulses are in error, the condition is described as a *2-bit error* or a *3-bit error*, and so on. When errors occur very close together (i.e. separated by relatively few correct bits), they are known as *burst errors* or an *error burst*. ⇨ *error correction* and *cyclic redundancy check*.

transmission header A part of the *header information* attached to a *packet* and providing *information* to allow the *access path control* system to determine the correct *route*.

transmission line A means to connect two *terminals* so that *messages* can be conveyed between them. Often terminals are connected by a two-way transmission line providing a *channel* to transmit in either direction. Transmission lines vary in physical characteristics, depending upon the

type of *message* traffic and the distances over which the *traffic* is to be conveyed.

transmission links Refers to the different physical elements which may make up the *transmission channels* from source to destination in a *telecommunications* system; for example, *local lines, local exchange, trunk lines, trunk exchange, satellite channel*. Generally speaking, these elements are not apparent to the *user* who operates a terminal to generate and receive a *message signal*.

transmission loss An impairment of a *message signal* due to the characteristics of the *transmission medium*, which causes a reduction in signal strength as a proportion of the energy is absorbed in the *transmission channel*.

transmission medium The medium used for the transmission of *message signals* in a *telecommunications* system; e.g. *twisted wire pairs*. A transmission medium is classified as a *bounded medium* or an *unbounded medium*. Examples of a bounded medium include twisted wire pairs, *co-axial tubes, waveguides* and *optical fibres*. Radio is an example of an unbounded medium.

The characteristics of a transmission medium define the quality of the *transmission channel*, which can be described in terms of its *bandwidth* and susceptibility to impairment of the message signals (e.g. impairment by *attenuation*).

transmission path A *communication channel* established for a particular *call* and pertaining to the specific resources utilized to effect the call; including physical *circuits* and *logical channels* where applicable.

transmission systems The term 'transmission system' refers to the physical media used to transmit *message signals* between locations. The simplest *transmission link* consists of a pair of wires connecting two *terminals*. This form of transmission link is used to connect *local lines* to an *exchange*, but it would be an expensive way of connecting several terminals over a *route* of more than, say, 5 kilometres. It is cheaper to use a method of transmission which permits the *multiplexing* of a number of separate *channels* on to the same *transmission medium*.

The equipment needed to multiplex and demultiplex the individual *messages* on to a *transmission channel* of high capacity is expensive, but it provides great savings in the construction and maintenance of individual lines, and of the ducts and cableways required to support these lines.

In multiplexing transmission systems, it is usual to provide a pair of wires, or a *co-axial tube*, in each direction, so each transmission system uses two pairs or tubes. Using this method a *co-axial cable*, which may contain 30 separate tubes, can support 15 transmission systems. The full *bandwidth* of each tube is used for transmission in any one direction and could support several hundred separate channels (⇨ article on *multiplexing*).

The performance of a transmission system is determined by the various physical elements in the overall system including *multiplexors* and *demultiplexors*, *transmission medium* (e.g. co-axial tubes), and the quality and physical spacing of *repeaters* along the transmission link.

Transmission systems for handling *analog voice signals* normally use *frequency division multiplexing (FDM)* and, for *digital signals, time division multiplexing (TDM)*. (It should be noted, however, that *pulse code modulation (PCM)* techniques are increasingly used to convert analog to digital signals in part of the transmission system.) Of course, a transmission system need not utilize only terrestrial media, such as cables, but could contain radio *satellite* links or line-of-sight *microwave* radio links.

transmission test set A device used by engineers to assist in *fault diagnosis* in *data communication systems*. Usually the test set is a portable item which can be connected to a *line* by an engineer or, in some cases, by non-skilled personnel. The test set can generate a number of *signals* which are of

a predetermined form, according to the speed and nature of the *circuit* to be tested. These signals are transmitted to the line and can be looped back to the test set so that the received pattern of signals can be compared with the internally generated pattern. A count of *bit errors* is recorded and displayed on the test set to show the quality and performance of that particular circuit.

transmit-to-receive crosstalk A form of *crosstalk* which arises on a *four-wire circuit* in which a *signal* transmitted in one direction is inducted on to the opposite *channel*. The crosstalk arising should be less than 1 to 40,000 relative to the transmitted signal, and at this level it is insignificant for *voice* and *data traffic*.

transmitter clock In *synchronous transmission*, a device which generates the *time intervals* to control the precise timing of *bits* passing from a *terminal* to a *line*. ⇨ *receiver clock*.

transmitting aerial A device used in *radio transmission* to radiate energy into the atmosphere as *electromagnetic waves*; it is also known as an *antenna*. The *radio signals* conveyed by this means are detected by a *receiving aerial*. Both transmitting and receiving aerials may be directional or non-directional; i.e. designed to transmit and receive in one particular direction or in any direction. Aerials are sensitive to particular *frequency bands* and this is determined largely by the physical dimensions of the aerial.

transmitting earth station A system designed to transmit *radio signals* to a *communications satellite* in earth orbit, in order that the signals may be reflected to another point on the earth's surface and be received by a *receiving earth station*.

transmitting station
1. Any *terminal*, or a *concentrator*, controlling a group of terminals which at a

particular instant in time is transmitting *signals* to a communication *network*.

2. A *radio transmitter* designed to propagate *radio waves*.

transmitting terminal Any *terminal* attached to a communications system and operating at a particular instant in time to transmit *signals* to the system.

Transpac Name given to a *packet switching network* developed and installed as a public service system in France.

transparency A concept in *computers* and communications in which two devices or logical units may communicate with each other without being aware of intervening equipment and *software*. This intervening equipment is said to maintain a transparent *interface*; i.e. it can be replaced by any more powerful or less powerful set of equipment, provided the new equipment maintains the same *logical* and *electrical interface*.

transposition This relates to the pattern of interconnection between groups of interconnecting *links* in an *exchange*; the level of interconnection may allow every group of *crosspoints* to be connected to every other group, giving *full availability* transposition. The purpose of *grouping* through successive stages of *switching* is to have a low probability of *call blocking* whilst reducing the number of physical crosspoints required. ⇨ *concentrator*, *distributor*, *expandor* and *grouping*.

tree network A *network* in which there is a hierarchy of *nodes* providing *control* and communication, and which can be represented on paper as the inverse of a family tree. The apex of the network, or top of the tree, represents the primary control for the network, but certain levels of control may be delegated down to intermediate branches. ⇨ *network topologies*.

triad An arrangement of three phosphor dots to represent a *picture element* on a

screen of a *colour television receiver*. Each triad includes a red, green and blue phosphor dot from which the colours of a televised scene can be composed.

trunk ⇨ *trunk circuit*.

trunk circuit This is the *transmission path* between *trunk exchanges* within the *main* (or *trunk*) *network* of a communication system. Such paths normally have to carry high *traffic volumes* over long distances, and some form of *multiplexing* takes place. ⇨ general article on *transmission systems*, and *exchange hierarchy*.

Often referred to as a *trunk link*, *trunk route* or simply a *trunk*.

trunk exchange An *exchange* which provides for the *switching* of *calls* which are not *local calls*, but which require to be connected with distant *subscribers* via *trunk circuits* and other *trunk exchanges*. Also referred to in the USA as *toll office*. ⇨ *exchange hierarchy*.

trunk line Any *line* forming part of the *trunk network* (i.e. one which is used to connect *trunk exchanges* and thus designed to carry *multiplexed traffic*). Usually consists of a physical medium of relatively *broad bandwidth*, and thus may carry hundreds or thousands of *logical channels*.

trunk link ⇨ *trunk circuit*.

trunk network That part of a public *telephone system* (or other communication systems) which consists of the major *transmission paths* and *exchanges* which provide for the interlinking of *subscribers* in the total *network*. It is usually defined to exclude the *local network* which incorporates the *local exchanges* and local *subscriber lines*. Also known as the *toll network* or *main network*. ⇨ general article on *exchange hierarchy*.

trunk route ⇨ *trunk circuit*.

TST network ⇨ *time-space-time network*.

tuned circuit A *circuit* designed to respond to a particular frequency to which it has been tuned by varying characteristics of a component forming part of the circuit; e.g. a circuit for selecting *signals* related to *radio transmission*.

tuner An element in a *radio* or *television receiver* which enables the *user* to select a particular broadcast *channel*.

turn-around time The time taken to reverse the direction of *data flow*. An expression used when referring to a *half-duplex* channel in which the directional flow of data can be reversed.

TV cable system A system in which *signals* to be detected and displayed by a *television receiver* are propagated along a *co-axial cable*. These techniques are applied to geographic areas of poor radio reception and to provide programmes of special or local interest which can be propagated over a cable without interfering with other *radio transmissions*. Cable TV systems are also used in industrial *applications* to provide remote inspection of processes or conditions.

TV receiver ⇨ *television receiver*.

twisted wire pairs The simplest and most common *transmission medium* used in *telephony*, in which the *transmission channel* consists of a single wire with an earth return. They are used for links between individual telephone *terminals* and *local exchanges*. The cable pairs are twisted to reduce the effect of *crosstalk* with other pairs which may be carried in the same cable or duct. Twisted pairs have limited application for *broadband multiplexed channels* using *analog signals*, because at higher frequencies *attenuation* and crosstalk are severe. By using *repeaters* to re-generate the *message signals* at frequent intervals, twisted wire pairs have been used with frequencies up to 500 kHz.

2-bit error An error occurring in a *data transmission* sequence in which two consecutive *bits* are incorrectly received. ⇨ *transmission errors*.

two condition code Any *code* used to represent *binary information* consisting of the values 0 and 1, and to be contrasted with a code in which *multi-state signalling* is used to represent more than just two values (for example, to represent 00, 01, 10 and 11). ⇨ *two-state signalling*.

two-state signalling A method of transmitting *information* in which only two voltage values are used serially to represent the *binary numbers* 0 and 1. Each *unit signal element* has a predefined duration, and the frequency of transmission occurs as an alternation between one voltage value and the other. If more than two voltage levels are used at the same frequency, the *data signalling rate* is higher; e.g. using 4 voltage levels, the binary numbers 01, 00, 10 and 11 can be represented.

two-way call A *call* in which *users* at each end of the *circuit* can act to send or receive *messages* simultaneously. For example, in a *telephone* call, the *subscribers* are linked by a two-way *transmission channel* over which they both can speak and listen.

two-wire channel A *transmission link* consisting of just a single pair of wires, which is most commonly used for *local lines* connecting telephone *subscribers* to a *local exchange*. It may also be used as a *trunk link* over short distances up to 2 kilometres. A two-wire channel is normally converted to a *four-wire channel* to facilitate *amplification* and to permit *multiplexing*.

U

UHF A range of electromagnetic frequencies used in *broadcasting*. ⇨ *ultra high frequency*.

UIT ⇨ *Union Internationale des Télécommunications*.

ultra high frequency *Radio waves* in the range from 300 MHz to 3 GHz which can be propagated as *space waves* over distances up to 200 kilometres. Used primarily for television *broadcasts* and, because of the limited range of propagation, relatively free from interference by distant transmissions in the same *band*.

unbalanced double-current interchange circuit An *interface* circuit between a *data terminal* (*DTE*) and a *modem* having certain electrical characteristics as defined in the *CCITT* recommendation V10, concerned with *data transmission* over telephone *circuits*. (⇨ *V series*.)

unbounded medium A *transmission medium* in which the *signal* is radiated in all directions, rather than directed along a physical medium such as a cable or an *optical fibre* link. The latter are examples of a *bounded medium*; a radio *broadcast* transmitted through the atmosphere is an example of an unbounded medium.

uniform quantization A *quantization* method in which the intervals of quantization are equal.

unintelligible crosstalk *Crosstalk* which results in intrusive but unintelligible *speech signals* being transferred from one *circuit* to another adjacent circuit.

Union Internationale des Télécommunications (UIT) This organization was originally founded in 1865 as the International Telegraph Union (*ITU*) and later

it became the *International Telecommunication Union*. In 1947 it became a special agency of the United Nations. Its main objectives are:
1. the promotion of international cooperation for efficient use of *telecommunication* services;
2. creation of widest possible public access to communication facilities;
3. harmonization of activities between national interests in this field.

The membership of the UIT at present numbers some 154 countries and its headquarters are in Geneva, Switzerland. ⇨ *Comité Consultatif International Télégraphique et Téléphonique (CCITT)*.

unit separator A special *code* under the general classification known as *information separators* and used in a *data communication code* to denote the boundary between *fields* in *binary coded* form.

unit signal element Refers to the *time interval* allocated to a basic unit of *information* in a *signal* used to represent binary information; i.e. to represent *binary digits* 0 and 1. The structure of the *signal element* will depend on the method of transmission – it may be a single pulse of defined duration or, as in the case of *frequency shift keying* and in *variable amplitude modulation*, may be a number of contiguous cycles at a particular frequency or *amplitude*, and used to represent a signal value.

unobtainable tone A *signal* heard by the *user* of a *telephone* when the number the person is calling is out of service.

upper sideband The process known as *modulation* (imprinting a *message signal* on to a *carrier wave*) generates a series of harmonic frequencies known as the *spectrum envelope*. For each *carrier* frequency, the principal outputs generated are two *sideband* signals, equally displaced in frequency. One, the *lower sideband*, is below the frequency of the carrier; the other is above and is known as the *upper sideband.* ⇨ *modulation.*

US ⇨ *unit separator.*

usage charge A charge made to a *user* of a communications service as part of the *tariff* for use of the service. The usage charge is based upon the actual utilization made by the *subscriber* and may include the number of *calls*, duration, volume, distance, and time segment in which the calls are made. Contrast with *standing charge*, and ⇨ *volume sensitive tariff.*

user A person sending or receiving *message signals* in a *telecommunication* system. A user utilizes a *terminal* to provide inputs or obtain outputs. Although it is common practice to consider the users to be outside the system, it is important that systems are designed to meet *user characteristics.*

user application In a *computer* system, this expression refers to the programs and procedures which define the particular job being performed for a group of *users*, as distinct from other general-purpose programs which may support the user application, such as the *operating system* or *teleprocessing monitor.*

user characteristics It is extremely important in designing *telecommunication* systems to evaluate and allow for the requirements of *users*.

For example, a *telephone network* requires to provide a level of reception satis-

factory to normal users wishing to carry out a two-way *conversation.* Telephone administrations carry out tests of user perception to identify a level of reception which is satisfactory, without incurring unnecessary cost or complexity in the system.

user data That part of a *message* (or *packet*) which contains *information* for the end user's *application*, rather than *control* or *header information* used to regulate transmission.

user data packet A unit of *information* forming part of a *message* transmitted from one *user* to another in a *packet switching network.*

user data signalling rate The rate at which a *terminal* can receive *information*, usually expressed in *bits* per second. This may not be exactly the same rate as information received by any *modem* associated with the *terminal* since, in some forms of *data transmission*, *envelopes* of *control information* pass between distant modems, but only higher-level user information within the envelopes is passed to and from the terminals.

user dialling state A *state* that exists when a *subscriber* is calling another *terminal* and has started (but not completed) the dialling process.

user node A point in a *network* at which a user *terminal* is stationed and used to distinguish from a *node* at which *control* functions, such as *switching* or *database* retrieval/updating operations, take place.

173

V

V series A series of recommendations by the *CCITT* which govern *data transmission* over telephone *circuits*. At the time of writing, the main recommendations are identified from V1 to V57. They are summarized below (compare also with *X series*):

V1 – provides equivalence between *binary notation* and the significance of a *two-condition code*

V2 – specifies power levels for data transmission over a telephone circuit

V3 – defines the *data communication code* known as *international alphabet No. 5*

V4 – defines the structure of *signals* to represent the international alphabet No. 5 for data transmission over public *telephone networks*

V5 – the standardization of *modulation rates* for *synchronous data transmission* and the associated *data signalling rates* on *public switched telephone circuits*

V6 – the standardization of modulation rates and data signalling rates for *synchronous operation* on *leased circuits*

V10 – a recommendation for the electrical requirements for *unbalanced double-current interchange circuits* for use with integrated circuits in *data transmission*

V11 – a recommendation for the electrical requirements for *balanced double-current interchange circuits* for use with integrated circuits in *data communication*

V13 – the definition of a simulator for an *answer-back unit*

V15 – requirements for the use of *acoustic couplers*

V16 – recommendations for special *modems* to transmit *analog signals* associated with the medical field

V19 – requirements for modems for *parallel data transmission* using signal frequencies relevant to telephone networks

V20 – a universal standard for modems for parallel data transmission over the *switched telephone network*

V21 – a standard for a 200-*baud* modem for use over the switched telephone network

V22 – a standard for a modem operating at 1200 bps in *full duplex* on a *2-wire circuit* over the switched telephone network

V23 – a standard for a modem operating at 600/1200 bps over the switched telephone network

V24 – a most important recommendation, since it is a standard adopted internationally by most *computer* manufacturers at the *hardware* level. It does not, however, govern the *software interface* which defines the method for controlling information flow across the V24 *interface*. This recommendation does provide an important hardware interface covering functions controlled, standard pin positions for connections and electrical characteristics.

Over 40 separate *control circuits* are defined and most *data terminal equipment* (*DTE*) as supplied by computer and communications companies will comply with a subset of this specification. The V24 recommendation allows for: local connections of terminals to a computer over a cable length of 50 feet; operation with modems over switched circuits; and operation with modems over *leased lines*. The recommendation consists of detailed definitions for *interchange signals* between the data terminal

(DTE) and the modem (DCE).

V25 – a recommendation for *automatic calling* of *automatic answering* equipment on the public switched telephone network

V26 – a standard for a modem operating at 2400 bps on a *four-wire circuit* for *point-to-point* connection

V26 bis – a standard for a modem operating at 2400/1200 bps over the switched telephone network

V27 – a standard for a modem operating at 4800 bps over a leased circuit

V27 bis – a standard for a modem operating at 2400/4800 bps for use on leased circuits with *adaptive equalizers*

V27 ter – a standard for a modem operating at 2400/4800 bps for use in the public switched telephone network

V28 – a recommendation for the electrical characteristics for unbalanced double-current interchange circuits

V29 – a standard for a modem operating at 9600 bps over a leased circuit

V31 – a recommendation for the electrical characteristics for *single current interchange circuits* controlled by closure of physical contacts

V35 – a recommendation for data transmission at 48 *kilobits* per second over circuits in the *group* from 60 to 108 kHz

V36 – a standard for modems used in *synchronous transmission* over circuits in the group from 60 to 108 kHz

V40 – the specification of error indications using electro-mechanical equipment

V41 – the specification of a *code independent* error control system

V50 – a definition of recommended limits for quality of circuits in data transmission

V51 – relates to the method of organizing maintenance for telephone circuits used in data transmission

V52 – a specification of the characteristics of equipment used for measuring *distortion* and *error rates* in data transmission over telephone circuits

V53 – specification of limits for the maintenance of telephone circuits used in data transmission

V54 – the definition of devices used for conducting *loop tests* on modems

V55 – specification of equipment for measuring *impulsive noise* on telephone circuits

V56 – a series of comparative tests for modems

V57 – the definition of a data test set for measuring the performance of systems at high data signalling rates

variable amplitude modulation A form of *amplitude modulation* in which binary information (i.e. 0 and 1 *bits*) are each represented by predetermined levels of *amplitude*. Several contiguous frequency cycles at a particular level represent 1 and at another level represent 0. ⇨ *modulation* and *unit signal element*.

variation of insertion loss The *insertion loss* experienced in a *circuit* is not usually constant but varies with time. In specifying the *quality of a circuit*, it is usual to state the possible variation of insertion loss. For example, an 800 Hz *signal* may produce an insertion loss of +10 *decibels* (*dB*), but it may vary between +7 dB and +13 dB.

vertical resolution A definition of the number of discrete *picture elements* which can be resolved along a vertical line drawn upon a display screen; e.g. in the British 625-line television system, 402 picture elements; in the USA 525-line system, 340 picture elements. ⇨ *resolution*.

vertical tabulation A function represented as a special *format effector* in a *data communication code*, serving as an instruction to move a *print mechanism* or the *cursor* of a *visual display unit* to the next predetermined position down the page of a document.

175

very high frequency (VHF) *Radio waves*, in the range from 30 MHz to 300 MHz, which can be propagated as *space waves* over distances up to 200 kilometres. Used primarily for high-quality radio *broadcasts* using *frequency modulation* techniques, where a relatively wide radio frequency *bandwidth* can be used for a much narrower *audio frequency* signal. This practice offers a high *signal-to-noise ratio* and, because of the limited range of VHF signals, there is relatively little interference from distant *stations* using the same part of the *frequency band*.

The VHF spectrum is also used for television broadcasts.

very large scale integration (VLSI) The application of *micro-electronic technology* to the integration of tens of thousands of components on to a single *silicon chip* of small physical size. Such devices can be designed, manufactured and tested by automated methods.

vestigial sideband A portion of a *sideband* which remains after *amplitude modulation* of a *carrier wave* using *vestigial sideband modulation* (*VSB*) techniques.

vestigial sideband modulation (VSB) A form of *modulation* used in television systems to remove almost the whole of one of the *sidebands* plus the *carrier wave*; e.g. the complete *upper sideband* will be transmitted but only part of the *lower sideband*. This technique is adopted to economize in the power and *bandwidth* required in transmission of the television signal, whilst maintaining a waveform which can be handled without unnecessary complexity and cost in each *TV receiver*.

VFT ⇨ *voice frequency telegraph.*

VHF A range of electromagnetic frequencies used in *broadcasting*. ⇨ *very high frequency.*

video amplifier A device used to increase the magnitude of a *video signal* received

from a television camera, prior to combining the *signal* with *sync pulses* and prior to modulating the signal for *radio transmission*. Also used to denote an *amplifier* in a *television receiver* used to increase the power of a detected video signal before it is delivered to the *cathode ray tube* for display.

video bandwidth This refers to the *bandwidth* required to carry a television picture, and depends mainly upon the amount of detail (i.e. *horizontal* and *vertical resolutions*) in the particular television system. This detail is given by the number of horizontal lines which form the *raster* and the number of *picture elements* along each horizontal line. In practice, some *line scans* take place during *field flyback*, when the *scanning beam* is blanked so the number of lines is multiplied by an appropriate factor less than unity (e.g. 0·92), in performing a calculation to derive the number of picture elements. In Britain where a 625-line system is used, the actual bandwidth required is 5·5 MHz. In France where a 625-line system is also used, a 6 MHz bandwidth is needed, because the French system produces a horizontal resolution which is greater than the vertical resolution.

video signal When a television picture is created by means of a camera focused on a scene, the image of the scene is formed on a light-sensitive surface inside the television camera. This establishes an electrical charge pattern which is related to the light coming from objects in the scene. This pattern is scanned by an *electron beam* inside the camera to create a video signal.

The scanning operation is achieved by *magnetic deflection coils* to produce horizontal and vertical movements of the beam. The pattern traced by the beam is a zig-zag movement across and down the screen, known as a *raster*. When the beam has been passed from left to right across the screen, it traverses rapidly back to the other side. This return motion is known as *flyback*.

The horizontal movement of the raster is known as a *line scan* and the vertical movement as a *field scan*.

The flyback movement after each line scan movement is known as *line flyback*. The return of the raster from the bottom to the top of the screen is known as *field flyback*.

The more lines occurring in the raster, the higher the definition of the image produced; i.e. a greater degree of detail can be captured from the original scene. The standards adopted vary from one country to another, but in Western Europe 625 lines are used in a television picture, and in North America 525 lines. The *CCIR* recommendation is for 625 lines.

The reproduction of a television picture in a *TV receiver* reverses the process described above. In this case, a beam is used to trace out the raster pattern on the screen of the receiver, and the beam current inside the *cathode ray tube* of the receiver is made to vary in sympathy with the video signal received from the *television transmitter*.

The beam is directed to strike luminous materials (phosphors) on the surface of the screen and to create patterns of light which relate to the video signal and thus relate to the light pattern emanating from the original scene.

Throughout this process, the camera scan must be synchronized with the scan of all the receivers and therefore *synchronizing pulses* are generated and transmitted during the *flyback periods* to control the beginning of each line and field scan movement. The *visual message signal*, therefore, consists of the video signal plus *sync pulses*; these two elements, when combined, are known as the *composite video signal*. ⇨ *television transmission signal*.

video traffic The video component of a television transmission, and normally carried on a television circuit which provides a unidirectional path of suitable *bandwidth*.

videotex adaptor A device which *interfaces* a standard *television receiver* to a telephone *line* so that it can be used to display *information* from a *videotex system*. Usually external to the television, it consists of a *modem, frame store, decoder* and a lead to connect to the *VHF* or *RGB* socket of the television.

videotex decoder A micro-electronic device fitted to a *television receiver* to enable it to receive, display and send *text* and *graphics* to a *telephone* line. ⇨ *videotex system*.

videotex standards Standards required to create compatibility between different national *videotex systems* and, for example, to allow for interaction between *videotex terminals* and *databases* in different countries. The standards cover: (a) *standard code sets* (*character sets*), (b) *extension codes*, (c) *composition coding*, (d) *graphic display standards*, (e) *dynamically redefinable character sets*.

The standard code sets include code set *C0* which provides definition of *cursor control* or *format effectors*, and *transmission control* characters. This is known as the basic control set or primary control set.

The C0 set (also known as primary *character set*) includes the basic set of *alphanumeric* characters.

The extension codes define how the basic sets of characters are to be extended; e.g. a supplementary set of *control characters* can be invoked by use of an *escape code* into an *extended control set* known as C1. Also, the *G1 code set* contains definition of the *mosaic graphics* set, and *G2 code set* contains special symbols and diacritical marks used in different languages.

However, the number of Latin-based character sets requiring diacritical marks is high, and to create devices capable of responding efficiently to all characters requires a technique known as composition coding. In this system, the character code and the diacritical mark are transmitted separately and then combined in the *terminal* for display. The recommendation prepared by *CCITT* for composition coding requires a few extra characters to represent all Latin character sets. It does, however, lead to extra logic and slightly more *page store* in each terminal.

Display standards cover two major techniques: (a) *alphamosaic coding*, (b) *alphageometric coding*. The first (a) allows *graphics* of *low resolution* to be displayed where pictures are formed of *graphic characters* made up by dividing each character space into a matrix of 2×3 elements. The pictures thus created are made up of combinations of up to 63 special graphic characters. This system enables graphics to be produced using cheap terminals.

The alphageometric coding system enables graphics of much *higher resolution* to be composed, including line drawings composed of arcs, points and polygons. The early geometric systems produce a resolution of 320×240 *picture elements* (*picels*). To utilize such a system efficiently in a *public videotex* service requires a *code compression* technique to transmit *picture description instructions* (*PDIs*) which are transmitted to the *line* and interpreted in the terminal. Without such a technique, every picel would have to be individually addressed down the line, which would provide unacceptable transmission delays in a *telephone network*.

A further refinement for alphamosaic systems is represented by the concepts inherent in *dynamically redefinable character sets* (*DRCS*). In this system, extra character sets are downloaded into the memory of the terminal from the videotex database system as required. Thus, a terminal can be programmed to display different language character sets or special symbols automatically before entering a particular sequence of *frames*. This standard is under consideration by the CCITT.

Also being considered are standards for storing, transmitting and displaying photographic images within a videotex network, and a prototype system known as *Picture Prestel* has been demonstrated

publicly by *British Telecom*.

These standards issues have a significant impact on the future development of videotex systems. It seems likely that a hierarchy of standards must exist, to allow for the evolution of a mass communication system based upon cheap technology in the terminal, which has to be produced by the TV manufacturers and by the makers of semi-conductor components needed in the *videotex decoders*.

videotex system The generic name given to an information system which allows for the collection and retrieval of *information* through *terminals* which are specially adapted *television receivers*, and are able to interact with a *database* stored in a central *computer*, using the *public switched telephone network* as the basic communication medium.

In Britain, the generic name *viewdata* is used; the public service offered by *British Telecom* has been called *Prestel*. This system was the world's first *public videotex* service and, after a public trial lasting some years, it was formally launched in 1980.

Systems launched or under development in other countries include: *Bildschirmtext* – West Germany; *Viditel* – Holland; *Teletel* – France; *Telidon* – Canada; *Captains* – Japan.

The concept of the service is that suppliers of information (known as *information providers* (*IPs*)) are able to offer information for sale to members of the business or domestic community. IPs may also provide information at no charge; e.g. to advertise products and services.

Users of the system requiring access to information are able to dial a connection to the *computer* which stores the *database* of IP information. They are then guided by an index to select those *pages* of information (known as *frames*) that they require.

Use of the service may result in three separate charges to the user: (a) the telephone call (usually at *local call rates*); (b) the computer connection; (c) the charge for each page examined.

The service is usually run by a *tele-communications* authority drawing their revenue primarily from the use of the *telephone network* and the computer service. The authority may also collect the *page charge* but normally pass this revenue on to the information providers.

Such systems can also be used to collect information from users; e.g. goods and services can be requested by a user who enters information through a *keyboard*. This process of *data collection* is associated with preset pages of information (known as *response frames*) designed by the IPs as order forms to be completed by users.

Public videotex systems usually incorporate very large and diverse databases maintained by (maybe) hundreds of information providers. The systems presently in use have thousands of terminals, and it is argued that by 1990 the *videotex terminal* will be as ubiquitous as the telephone today.

Private videotex systems are also being implemented, and the British systems company known as *Aregon* has pioneered the installation of such systems in commercial organizations wishing to use the system for internal communication between distributed offices, and from principals to their agents and customers.

A key feature of a videotex system is the simple user protocol which requires only a few minutes' training. Further details of the types of terminal available are described under the heading *videotex terminals*. Usually the terminal has a display format of 40 columns by 24 lines. At the time of writing, most installed systems transmit information at 1,200 *bauds*, and provide *text* and limited *graphics* including up to 7 colours in a single display. The *send channel* from the user to the computer has to handle only simple requests and usually operates at 75 bauds.

The major components in a videotex system are: (a) *digital computer* to store the database; (b) *telephone system* to provide the *network*; (c) 1200/75 baud *modem* to connect terminals to the network; (d)

simple user terminals to access and display pages; and (e) *editing terminals* to allow information providers to create *pages*.

For further definitions associated with this concept, ⇨ *videotex terminal, videotex standards, alphageometric coding, alpha-mosaic coding, character generator, composition coding, dynamically redefinable character sets, parallel attribute coding, Picture Prestel* and *serial attribute coding*.

Videotex systems should be compared with *teletext* broadcast systems which use *radio waves* instead of the telephone network. Teletext provides a cheap mass-communication system in which user interaction is limited, whereas videotex provides for user interaction with a large database at any time with uniform access time.

videotex terminal A terminal designed to receive and display *signals* from a *videotex system* and to enable *users* to retrieve or input *information* to such a system. The concept of videotex is based upon the use of a domestic *television receiver* as a *terminal*, able to communicate with a *database* over the *public switched telephone network*. This requires the TV receiver to be modified to include a special *videotex decoder* and a *modem*. This modification may be made inside the receiver during its manufacture or by an external *videotex adaptor* which plugs into the *UHF* or *RGB* input sockets to a standard TV. The additional components required to convert a standard TV receiver include the following: (a) *modem* – which converts *analog signals* received over the telephone network to *digital signals* for the *decoder*; (b) *line isolator* – which isolates the telephone network from high voltages and extraneous signals from the TV receiver; (c) *memory unit* – which stores either one or more pages for display, and enables pages to be created by a *character generator* for display; e.g. in a 6 × 10 dot matrix for each character; (d) *processing logic* – to interpret incoming signals and create pages for the memory unit; (e) *keyboard* or *keypad* – to enable the user to issue commands to the system and to input *data*: (f) *automatic dialler* –

an optional feature to enable users to select the telephone number of the particular videotex service required; otherwise the telephone dial is used in the normal way.

In most countries there is commonality between the functional components of a *videotex decoder* and a *teletext decoder* which responds to *radio signals* rather than to signals from the *telephone system*.

The accepted display standard for videotex terminals consists basically of 24 rows of *characters* per *page* or *frame* and 40 characters per row. It is referred to as a *40 column display* which compares with the 65 column or *80 column display* more commonly found with *computer* terminals.

The videotex terminal will usually display up to 7 colours, with *text* and *graphics*, and a range of basic characters described under videotex standard *code sets*. There are many variations in videotex terminals. The first distinction to make is that there are: (a) standard *information retrieval* terminals used by users to request pages and to input data but not capable of editing functions to create pages, usually operated by a keypad, allowing selections 0 to 9 and a few special functions to be entered; and (b) *editing terminals* used by *information providers* to create pages and to maintain the *database*.

The standard information retrieval terminal may have an alphabetic keyboard to enable *messages* to be entered or *response frames* to be completed, but often it will have only a simple numeric keypad to enable choices to be indicated plus a few simple *control* functions. There is also a range of screen sizes for videotex terminals, ranging from the large domestic TV to be viewed at a distance, to small executive terminals to be positioned on a desk close to the viewer. Not all videotex terminals are capable of TV reception, but those intended for domestic use are usually used alternatively for TV viewing.

It is also important to note that there are different degrees of *resolution* for *graphic displays*. Often, alternative degrees of picture resolution are available on the same system. This topic is described in

more detail under *alphamosaic graphics* and *alphageometric graphics*. ⇨ *videotex standards*.

videotex traffic *Message signals* representing *characters* of *information* transmitted over the *public telephone network*, or a *packet network*, between *videotex terminals* and a central *videotex system*. The line protocol is based upon the *ASCII data communication code* for *text*.

Viditel The name given by the *PTT* in Holland to its *public videotex* trial, first introduced in Holland starting in 1980. ⇨ *videotex system*.

viewdata A generic term used in English-speaking countries to refer to a public information service which incorporates the use of modified *television receivers* to gain access to *information* stored in a *computer*, and transmitted to *users* over the *public switched telephone network*. For details, ⇨ the general article under the *CCITT* preferred term – *videotex system*.

viewdata adaptor ⇨ *videotex adaptor*.

viewdata terminal ⇨ *videotex terminal*.

virtual circuit In a *packet switching* system, *messages* share *links* between various *nodes* in the *data communications* network, usually using some form of *time division multiplexing* (*TDM*). Consequently, there is no *access path* associated with each *call* to provide an end-to-end connection for the duration of the call. Instead, messages are transmitted as a number of small *packets* of *binary information*, which contain *routing* and *address information*. These packets pass through the network, sharing certain paths with packets from other messages, and packets are reassembled (or *de-packetized*) on arrival at their *receiving station*.

The path through such a *network* for any particular message is referred to as a virtual circuit, and it can provide a *full duplex* link between the *high level control functions*

at the *transmitting* and *receiving terminals*.
⇨ *permanent virtual circuit.*

visible colour spectrum The spectrum of
electromagnetic waves which gives rise to
colours as perceived by humans, with a
span of colours and *wavelengths* including
those shown below:

colour	approximate wavelength n/m
deep red	730
red	670
orange	610
yellow	580
green	530
blue	460
violet	410

(1 nanometre (n/m) = 10^{-9} metres)

Humans cannot perceive colours
beyond this spectrum; e.g. (ultra violet
light around 380 n/m and *infra red light*
around 780 n/m.

The visible colour spectrum corresponds
to electromagnetic waves in the *frequency
spectrum* from 420 to 790 THz (1 THz =
10^{12} Hz).

visible light spectrum The range of *electromagnetic waves* which are visible to
humans, including the frequencies between
violet and deep red. This spectrum contains
all the frequencies which can be combined
to produce the various colours perceived
by humans. ⇨ *visible colour spectrum.*

visibility function This relates to the
average visual response of humans to
electromagnetic waves which varies with
frequency or *wavelength*. The human eye
can respond to wavelengths between those
corresponding to violet (400 nanometres
(n/m)) and deep red (760 n/m). However.
it cannot perceive ultra violet or *infra red
light*, which have wavelengths beyond this
spectrum. Also, the response varies within
the *visible colour spectrum*, and the average
human eye is far more responsive to light
in the yellow spectrum around 580 n/m.
(1 n/m = 10^{-9} metres.)

This visibility function has to be considered in all forms of *visual display* and

in the concepts of television. A source of
red light at 680 n/m would require 100
times more luminous power to invoke the
same response in a human as a source of
yellow/green light at 550 n/m.

The visibility function is also known as
the *photopic response* and is part of the
science of *photometry.*

vision Pertaining to the *video signals*
which originate from television cameras,
in which light reflected from an object is
detected and represented as an *electrical
signal.*

vision carrier In television broadcasting,
a *carrier wave* which is modulated using
a form of *amplitude modulation* to carry
the *video signal.* A separate carrier is used
for the *sound signal.* ⇨ *vestigial sideband
modulation* and *television transmission
signal.*

visual acuity This is a measure of the
amount of detail which can be detected
visually by humans. A formal definition
says that it is the reciprocal of the angle
α subtended at the eye by two small images
which can be separately identified. Acuity,
or sharpness of visual perception, has been
measured for people in tests conducted on
an international basis. It is dependent upon
a number of factors, such as shape of the
objects, the contrast with the background
and the intensity of light; but, given control
of these factors, the findings are that the
average viewer can just resolve 425 lines
on the surface of a TV screen at normal
viewing distance. In fact, the *raster* used
to generate such a display will need 607
lines. ⇨ *kell factor.*

visual display An input/output *terminal*
based upon the technology of the *cathode
ray tube* and used to display *data, text*
and *graphic information.*

visual message signal A component of a
television transmission signal which consists
of a *video signal* plus *synchronization pulses*
used to maintain *synchronization* between

a *TV camera* and a *TV receiver*. It does not include the *audio signal* which is associated with the visual message to form the complete television transmission signal. Also known as *composite video signal.* ⇨ *video signal.*

VLSI Abbreviation of *very large scale integration.* ⇨ micro-electronic technology.

voice Relating to the representation of human speech as *electrical signals*, as in *voice channel*, *voice frequency* or *voice traffic.*

voice analog signal An *electrical signal* which is a direct representation of an *audio signal* corresponding to human speech. ⇨ *analog electrical signal.*

voice band frequency A frequency occurring in the relatively narrow band used for carrying *signals* representing human speech, and usually in the range from 300 to 3400 Hz.

voice channel A communications *circuit* designed to carry *voice signals* and therefore designed to accept an *electrical analog* of an *audio frequency signal* corresponding to human speech. But note that today speech can be converted to *digital signals*, using *pulse code modulation* techniques.

voice circuit A *circuit* used to carry *signals* representing human speech and normally having a *bandwidth* of from 300 to 3400 Hz.

voice digitization Pertaining to the techniques used to convert *speech signals* into *digital signals* by means of *pulse code modulation*. Digitization can take place in the telephone *handset* or at the point of entry to a *digital exchange.*

voice frequency Relating to the range of *signals* used to convey a representation of human speech in a *telephone system*, and normally in the *bandwidth* from 300 to 3400 Hz.

voice frequency telegraph A *telegraph system* in which *signals* are transmitted at *audio frequencies* at all intermediate stages in the *network.*

voice packet switching Refers to the use of *packet switching networks* to carry speech information. Although such networks were originally (and are primarily) used for *data*, it is possible to transmit speech. Indeed, economic studies undertaken on behalf of the US Defense Department have shown significant cost benefits in the transmission of speech in this way. The *electrical analog signals* representing speech must first be digitized, using *pulse code modulation* techniques, then *packetized* in the entry *node* of the *packet network.* It has been shown that a network designed to present a nominal end-to-end *packet delay* of around 200 milliseconds will allow two people to enjoy a normal telephone *conversation.*

voice signal
 1. A sound energy signal corresponding to human speech.
 2. An *electrical analog signal* representing human speech and used to convey speech information as *electromagnetic waves*. Such a signal is usually within the *frequency band* from 0·3 to 3·4 kHz.

voice telephony The concepts and principles associated with the transmission of speech in a *telephone network.*

voice traffic *Traffic* corresponding to human speech and requiring a *channel* capable of handling an *analog signal* in the range from 0·3 to 3·4 kHz.

volume sensitive tariff A method of charging for the use of a *circuit* or *public data network* in which the charges are related to the volume of *traffic* a particular *subscriber* delivers to the system rather than distance. The unit of delivery is predetermined by the service operator and varies from service to service, but in *data transmission* systems a *data segment* of 64 *characters* may be defined as a minimum unit. Discounts or penalties may apply to *users* at certain *traffic volumes* within specified durations. ⇨ *tariff* and *usage charge*.

VSB Abbreviation of *vestigial sideband*.

VT ⇨ *vertical tabulation*.

VX Shorthand notation for *videotex*.

W

waveguides A *transmission medium* in which *radio waves* are propagated along hollow conducting tubes. The cross-sectional dimensions of the tube need to be equal to, or greater than, the *wavelength* of the radio waves. This means that waveguides are suitable for use with *microwaves*; e.g. in the range of 2 GHz to 11 GHz. ⇨ *optical waveguides*.

wavelength All *electromagnetic waves* travel through *space* at approximately the same speed: 300,000,000 metres per second. The wavelength of an electromagnetic signal is obtained by dividing this number by the signal frequency in cycles per second (Hertz).

Thus, the wavelength for a signal of 1,000,000 cycles per second (1 MHz) is 300 metres. The higher the frequency, the shorter the wavelength. ⇨ articles on *electromagnetic waves* and *bandwidth*.

white noise A form of interference which manifests itself as a hiss on telephone or radio channels, and is caused by the natural movement of electrons in *circuits* which varies with temperature. Also known as *gaussion noise* and *random noise*.

who are you A *character code* which occurs in a *data communications code* such as the *international alphabet No. 2*, and which has the purpose of invoking an auto-matic response from a *terminal* causing it to reply with a *call* sign.

wideband circuit A *circuit* of broad band-width; e.g. allowing *data transmission* at speeds of 48, 64 or 72 *kilobits* per second.

word A unit of *data* used in *computer* systems and operated upon by the computer as an entity. A word may consist of a defined number of *characters* and *bits*; e.g. 4 characters of 8 bits, making 32 bits per word. ⇨ *byte*.

word processor A system for the preparation and distribution of *messages* upon documents, based upon the principles of the conventional typewriter, but enhanced to provide more advanced facilities which are realized by *computer* techniques; e.g. editing features, automatic tabulating features, storage and filing capabilities, and automatic generation of type fonts, formats, and numbers of copies.

word processor networks A system in which *word processors* are interconnected by communications facilities to provide for the distribution of *information* recorded as documents within a word-processing environment. Such systems presage the evolution of integrated office systems based upon the concepts of *electronic mail* – the so-called office of the future.

X

X series (X1–X29) The *CCITT* has produced a number of standards, which are recommendations to *PTTs* and other communication authorities, to establish communications *interfaces* for users' *data terminal equipment* (*DTE*), and for *network interface units* or *modems* known as *data circuit terminating equipment* (*DCE*). The X series recommendations thus govern the attachment of *terminals* to *public data networks* (*PDNs*) and constitute a family of standards for *interfacing* terminals to networks. In principle, they recognize the growth of *bit-stream* oriented systems dealing in so-called *high level protocols*, and represent a major step in evolution from the long-standing *V series interfaces*, which dealt with the attachment of digital devices to *analog* facilities, represented by existing *telephone networks*.

Some of the more commonly used standards are listed below; a little more detail is provided under the headings *X21* and *X25*:

X1 – defines classes of service for international PDN *users*

X2 – defines international user facilities in a PDN

X3 – defines *packet assembly* and *disassembly*

X4 – defines the structure of *signals* to represent the *international alphabet No. 5* for *data transmission* over PDN

X20 – defines the interface for *start-stop transmission* services between DTE and *DCE*, including an extension (X20 bis) for *V21* compatible interfaces for such services on a PDN

X21 – defines the interface between DTE and DCE for *synchronous operation* on PDNs, including an extension (X21 bis) for terminals designed to interface to *V series* modems for synchronous opera-

tion. (But see more detailed definition under *X21*.)

X24 – definitions relating to *interchange circuits* between DTE and DCE on PDNs

X25 – defines the interface between DTE and DCE operating in packet switching mode in a PDN. (But see more detailed definition under *X25*.)

X28 – defines interface requirements for a non-intelligent start-stop mode terminal accessing the packet assembly and disassembly facility in a PDN

X29 – procedures for the exchange of *control information* and *user data* between a *PAD* and DTE capable of packet assembly and disassembly functions

There are a number of other X series recommendations and, for accurate up-to-date details, reference should be made to CCITT publications. The objective of the recommendations is to cover: (a) methods of terminal attachment; (b) *link control* procedures; and (c) *data transfer*, embracing: *leased circuits, switched circuits*, and *packet switching*. (Reference: CCITT Orange Book Vol. VIII. 2 Public Data Networks (Series X recommendations))

X21 A member of the family of *X series interfaces*, designed to replace the earlier *V24* and *V25* interface recommendations of the *CCITT*. It deals with the connection of *terminals* to *modems*, including *automatic calling* operations for *switched circuits* on synchronous *public data networks* (*PDNs*). The benefits of X21 over the equivalent *V series* include: (a) the number of *interchange circuits* and pin connections is greatly reduced, because many terminal-to-network *control* signals are represented as *code strings* rather than as discrete circuits; (b) the electrical characteristics defined for X21 incorporate a *physical interface* capable of supporting *large scale integration* micro-components (*LSI*), which can operate at very high *bit-rates* and over greater distances than the V series recommendation; (c) enhanced functions are available, including the ability to transfer more detailed call status information, relating to the *called terminal* and the network.

The X21 recommendation maintains *transparency* during the *data transfer* stage, and provides a distinct separation of *signalling* and data transfer.

X25 This recommendation of the *CCITT* is intended as a *protocol* for *packet switching networks*, and supporting the attachment of *intelligent terminals* as well as *communication controllers* and *host processors*. There are also some supporting recommendations (X3, X28 and X29) which relate to the support of *start-stop* terminals, as well as intelligent terminals.

The major emphasis with a *packet network* is that the *users data* must be formatted into relatively small units, known as *packets*. All packets delivered by user terminals to the network under the X25 *interface* must have *header information* to control the progress of the packet in the network. For terminals of the non-intelligent type, designed according to X3, X28 and X29 recommendations, the network creates the additional information and provides the function to create the correct packet formats.

An important aspect of X25 is that a user interacts indirectly with any remote terminal. The user interaction is, in fact, with the *node* through which it gains access to the network; i.e. it operates under a *node-to-network protocol* rather than an *end-to-end protocol*.

The X25 recommendation has a complex *high level function* which is structured in three levels: (a) physical and electrical – provides for the activation, maintenance and de-activation of *links* between the terminal (DTE) and the local *data circuit terminating equipment* or *network interface unit*; (b) *link access procedure* – provides *framing, synchronization, error detection* and *correction* between the terminal and the adjacent network node through which network access is obtained; (c) packet format and control – provides for the establishment, *data transfer* and clearing of *virtual circuits* between the terminal and the network via the adjacent network node.

The preferred link access procedure (*LAPB*) provides a *high level data link control*, which is a subset of the *ISO HDLC* protocol.

The recommendation of the packet format includes call request, *call accepted* and *call connected packets* which control the establishment and activation of virtual circuits, plus the definition of data packets which contain user information. The user data packet contains the following *control* and identification *data*: header format identifier, logical channel group number, *logical channel number*, packet receive sequence number, flag to indicate if more data is to follow, packet send sequence number, and *user data* (from 128 to 1024 *octets*).

The concept of logical channel number relates to the *channel* occupied by the *call* between the terminal and the network. The *channel identification* is assigned when the virtual circuit is set up. It should be noted in this respect that under X25 up to 4095 separate *logical channels* can be connected to a single input/output *port* on a *computer*.

It will also be noted that the users data packet does not contain the *called* and *calling addresses*; these are conveyed to the network as part of the *call request packet* originally used to set up the virtual call.

Z

zero loss A condition which arises when the *gain* in power in a *circuit* due to *amplification* is just sufficient to overcome the inevitable loss in power attributed to the *line* and associated operating equipment.

APPENDIXES

Appendix 1 – International Alphabet No. 5

This is a 7-bit code in which combinations of *bits* provide unique patterns to represent *characters*, numerals, special symbols and functions. Bits 1 to 4 form a basic group giving binary values from 0 to 15; e.g.

$$0 = 0000$$
$$7 = 0111$$
$$11 = 1011$$
$$15 = 1111$$

Bits 5 to 7 provide for 8 separate cases:

0 = 000	⎫	Transmission control functions
1 = 001	⎬	
2 = 010		Symbols
3 = 011		Numerals 0 to 9 plus symbols
4 = 100		Mainly upper-case letters A to O
5 = 101		Mainly upper-case letters P to Z
6 = 110		Mainly lower-case letters a to o
7 = 111		Mainly lower-case letters p to z

In *data transmission*, a *parity bit* is added to the *code*, making 8 bits in all.

The code does allow for national variations to cover the requirement for alphabet extensions or for diacritical signs. These positions have been left blank in the table overleaf (e.g. 4/0 and 5/14).

Column Row	0	1	2	3	4	5	6	7
0	NUL	DLE	SP	0		P		p
1	SOH	DC1	!	1	A	Q	a	q
2	STX	DC2	"	2	B	R	b	r
3	ETX	DC3	£ or #	3	C	S	c	s
4	EOT	DC4	$	4	D	T	d	t
5	ENQ	TC8	%	5	E	U	e	u
6	ACK	SYN	&	6	F	V	f	v
7	BEL	ETB	'	7	G	W	g	w
8	BS	CAN	(	8	H	X	h	x
9	HT	EM	)	9	I	Y	i	y
10	LF	SUB	*	:	J	Z	j	z
11	VT	ESC	+	;	K		k	
12	FF	FS	,	<	L		l	
13	CR	GS	-	=	M		m	
14	SO	RS	.	>	N		n	-
15	SI	US	/	?	O	-	o	DEL

Appendix 2 – Notation Used for Abbreviation of Frequency Ranges

Hz   Hertz   cycles per second
kHz   kiloHertz   thousand cycles per second
MHz   MegaHertz   million cycles per second
GHz   GigaHertz   thousand million cycles per second
THz   TeraHertz   million million cycles per second

Examples

50 kHz = 50,000 cps or 50×10^3 Hz
30 MHz = 30,000,000 cps or 30×10^6 Hz
300 GHz = 300,000,000,000 cps or 300×10^9 Hz
790 THz = 790,000,000,000,000 cps or 790×10^{12} Hz

Appendix 3 – Wavelength (2)

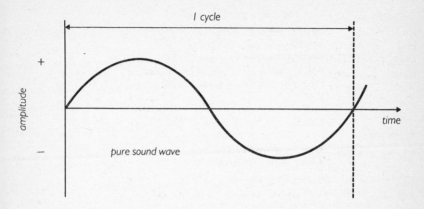

The velocity of *sound* in air is 340 metres/second. For a sound at a pitch or frequency of 1,000 cycles per second (1 kHz), the wavelength is 340/1000 or 0·34 metres.

The velocity of *electromagnetic waves* in space is 300,000,000 metres per second. Thus, for a *radio wave* of 30 MHz, the wavelength is $3 \times 10^8 \div 3 \times 10^7 = 10$ metres.

$$\text{Wavelength } (\lambda) = \frac{\text{velocity}}{\text{frequency}}$$

Appendix 4 – The Electromagnetic Spectrum

approximate wavelength (n/m)

gamma-rays	10^{-3}
X-rays	10^{-1}
ultra-violet rays	10
visible light	10^3
infra-red	10^4
heat	10^6
radar	10^7
radio	10^{11}

n/m   = nanometre(s)
1 n/m = 10^{-9} metres

Appendix 5 – Visible Light Spectrum

The visible light spectrum occurs between ultra-violet and infra-red frequency ranges and includes the following:

	wavelength (n/m)
violet	420
blue	460
blue-green	490
green	530
yellow-green	550
yellow	590
orange	620
red	660
deep red	740

Note: In each case a range of frequencies apply, and the table gives an approximate point in the range.

Appendix 6 – Bandwidth and Information

The basic relationship between bandwidth and information is given in the following illustration:

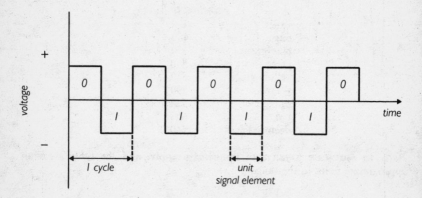

Assume maximum cycle frequency = 50 Hz
Then *data signalling rate*   = 100 *bits* per second (0 or 1)
and minimum frequency   = 0 Hz (i.e. all 0's or all 1's).
One hundred bits (or one hundred unit *signal elements*) per second can be achieved. This is referred to as 100 *bauds*.

In this example only 2 voltage levels are used but, by using 4 voltage levels, the *binary digits* 11, 00, 01 and 10 can be represented, thus increasing the data signalling rate to 200 bits per second at the same maximum frequency.

Appendix 7 – Basis of Amplitude Modulation

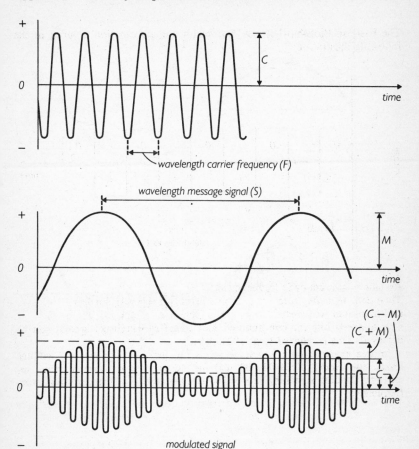

wavelength carrier frequency (F)

wavelength message signal (S)

modulated signal

The original carrier frequency F is modulated by a *message signal* S to create an *envelope* containing F and S. The envelope can provide:

the carrier frequency F
the message signal S.

The principles apply to *frequency modulation*, in which the message signal, instead of altering the *amplitude* of the *carrier signal*, causes the carrier to vary its frequency in sympathy with the frequency of the message signal.

Appendix 8 – Basis of Pulse Amplitude Modulation

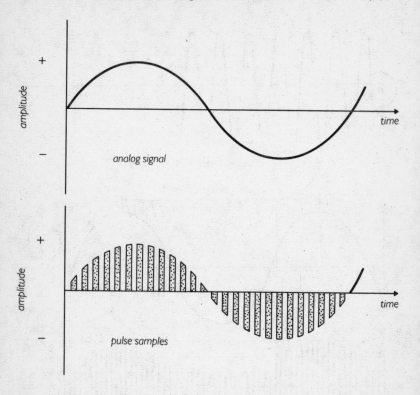

The *amplitude* of pulses is derived from the original *signal*. Approximately 8,000 *PAM* pulses per second are required to give a good reproduction of human speech over a 4 kHz *circuit*.

In practice, pulses of varying amplitude are not transmitted; instead, *binary coded information* is transmitted to represent the amplitude levels. Decoding logic at the receiving end reconverts the binary code to the amplitude signals and enables the original *analog signal* to be reconstructed.

Appendix 9 – Packet Switching Network

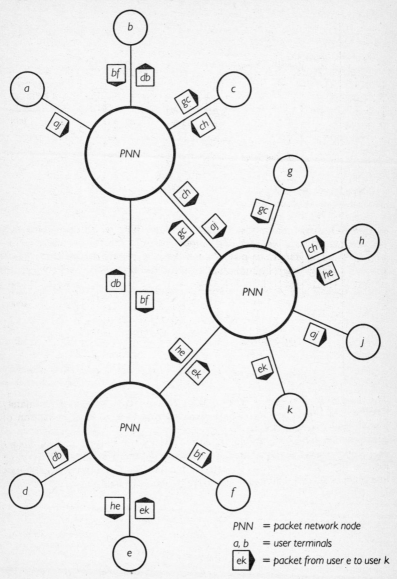

PNN   = packet network node
a, b   = user terminals
ek   = packet from user e to user k

Appendix 10 – ISO Architecture Showing Layered Protocols for X25 Usage

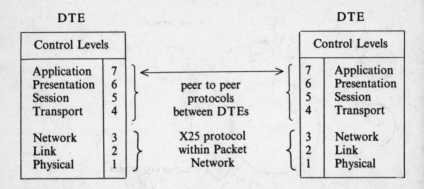

In the ISO provisional model, seven *control levels* are recommended for each *user* connected to a *network*. The model requires the further definition of *protocols* and *formats* for co-ordination between respective *control* layers. At the time of writing, international standards for the first three levels have been agreed.

Appendix 11 – Network Topologies

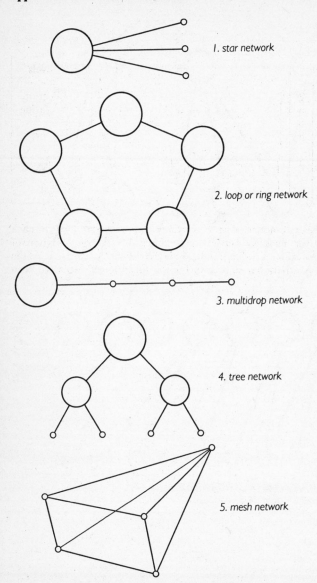

1. star network

2. loop or ring network

3. multidrop network

4. tree network

5. mesh network

Appendix 12 – Exchange Hierarchy in a Telephone Network

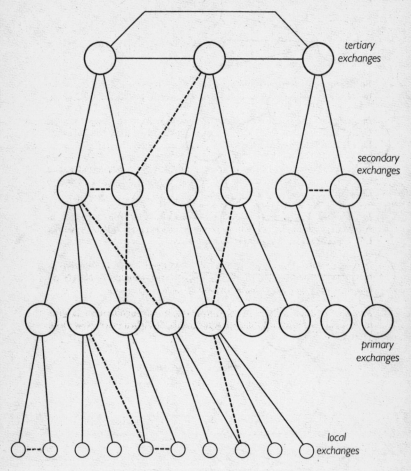

tertiary
exchanges

secondary
exchanges

primary
exchanges

local
exchanges

backbone routes indicated by thick lines,
alternative routes by broken lines

MORE ABOUT PENGUINS, PELICANS
AND PUFFINS

For further information about books available from Penguins please write to Dept EP, Penguin Books Ltd, Harmondsworth, Middlesex UB7 0DA.

In the U.S.A.: For a complete list of books available from Penguins in the United States write to Dept DG, Penguin Books, 299 Murray Hill Parkway, East Rutherford, New Jersey 07073.

In Canada: For a complete list of books available from Penguins in Canada write to Penguin Books Canada Ltd, 2801 John Street, Markham, Ontario L3R 1B4.

In Australia: For a complete list of books available from Penguins in Australia write to the Marketing Department, Penguin Books Australia Ltd, P.O. Box 257, Ringwood, Victoria 3134.

In New Zealand: For a complete list of books available from Penguins in New Zealand write to the Marketing Department, Penguin Books (N.Z.) Ltd, Private Bag, Takapuna, Auckland 9.

In India: For a complete list of books available from Penguins in India write to Penguin Overseas Ltd, 706 Eros Apartments, 56 Nehru Place, New Delhi 110019.

Also by John Graham

THE PENGUIN BOOK OF
PERSONAL COMPUTING

You want to use a computer in your home or office? How should
you choose it? What can it do?

The Penguin Book of Personal Computing forms a complete, illustrated, easy-to-follow course for everyone who is considering
buying a microcomputer, or already has access to one and wants
to know how it can be used. Compiled by expert computer writer
John Graham, this is an indispensable introduction to the fast-
growing world of personal computing. Topics covered include:
* how personal computers work
* attachments – keyboards, VDUs,
cassettes, disks, printers
* programming languages
* application programs from games to
business systems
* buying software, how to assess what's on
the market
* applications at home and in business
* networking, particularly over the
telephone

PLUS a guide and checklist to help you choose your system, a
bibliography of useful books and magazines, and a glossary of
terms.

A CHOICE OF
PELICANS AND PEREGRINES

A CHOICE OF
PELICANS AND PEREGRINES

☐ *A Question of Economics* **Peter Donaldson** £4.95

Twenty key issues – from the City and big business to trades unions – clarified and discussed by Peter Donaldson, author of *10 × Economics* and one of our greatest popularizers of economics.

☐ *Inside the Inner City* **Paul Harrison** £4.50

A report on urban poverty and conflict by the author of *Inside the Third World*. 'A major piece of evidence' – *Sunday Times*. 'A classic: it tells us what it is really like to be poor, and why' – *Time Out*

☐ *What Philosophy Is* **Anthony O'Hear** £3.95

What are human beings? How should people act? How do our thoughts and words relate to reality? Contemporary attitudes to these age-old questions are discussed in this new study, an eloquent and brilliant introduction to philosophy today.

☐ *The Arabs* **Peter Mansfield** £4.95

New Edition. 'Should be studied by anyone who wants to know about the Arab world and how the Arabs have become what they are today' – *Sunday Times*

☐ *Religion and the Rise of Capitalism*
R. H. Tawney £3.95

The classic study of religious thought of social and economic issues from the later middle ages to the early eighteenth century.

☐ *The Mathematical Experience*
Philip J. Davis and Reuben Hersh £6.95

Not since *Gödel, Escher, Bach* has such an entertaining book been written on the relationship of mathematics to the arts and sciences. 'It deserves to be read by everyone ... an instant classic' – *New Scientist*

THE PENGUIN ENGLISH DICTIONARY

The Penguin English Dictionary has been created specially for today's needs. It features:

* More entries than any other popularly priced dictionary
* Exceptionally clear and precise definitions
* For the first time in an equivalent dictionary, the internationally recognised IPA pronunciation system
* Emphasis on contemporary usage
* Extended coverage of both the spoken and the written word
* Scientific tables
* Technical words
* Informal and colloquial expressions
* Vocabulary most widely used *wherever* English is spoken
* Most commonly used abbreviations

It is twenty years since the publication of the last English dictionary by Penguin and the compilation of this entirely new *Penguin English Dictionary* is the result of a special collaboration between Longman, one of the world's leading dictionary publishers, and Penguin Books. The material is based entirely on the database of the acclaimed *Longman Dictionary of the English Language.*

1008 pages   051.139 3   £2.50 ☐

PENGUIN REFERENCE BOOKS

☐ *The Penguin Map of the World*   £2.50

Clear, colourful, crammed with information and fully up-to-date, this is a useful map to stick on your wall at home, at school or in the office.

☐ *The Penguin Map of Europe*   £2.95

Covers all land eastwards to the Urals, southwards to North Africa and up to Syria, Iraq and Iran * Scale = 1:5,500,000 * 4-colour artwork * Features main roads, railways, oil and gas pipelines, plus extra information including national flags, currencies and populations.

☐ *The Penguin Map of the British Isles*   £1.95

Including the Orkneys, the Shetlands, the Channel Islands and much of Normandy, this excellent map is ideal for planning routes and touring holidays, or as a study aid.

☐ *The Penguin Dictionary of Quotations*   £3.95

A treasure-trove of over 12,000 new gems and old favourites, from Aesop and Matthew Arnold to Xenophon and Zola.

☐ *The Penguin Dictionary of Art and Artists*   £3.95

Fifth Edition. 'A vast amount of information intelligently presented, carefully detailed, abreast of current thought and scholarship and easy to read' – *The Times Literary Supplement*

☐ *The Penguin Pocket Thesaurus*   £1.95

A pocket-sized version of Roget's classic, and an essential companion for all commuters, crossword addicts, students, journalists and the stuck-for-words.

PENGUIN REFERENCE BOOKS

☐ *The Penguin Dictionary of Troublesome Words*　£2.50

A witty, straightforward guide to the pitfalls and hotly disputed issues in standard written English, illustrated with examples and including a glossary of grammatical terms and an appendix on punctuation.

☐ *The Penguin Guide to the Law*　£7.50

This acclaimed reference book is designed for everyday use, and forms the most comprehensive handbook ever published on the law as it affects the individual.

☐ *The Penguin Dictionary of Religions*　£4.95

The rites, beliefs, gods and holy books of all the major religions throughout the world are covered in this book, which is illustrated with charts, maps and line drawings.

☐ *The Penguin Medical Encyclopedia*　£4.95

Covers the body and mind in sickness and in health, including drugs, surgery, history, institutions, medical vocabulary and many other aspects. Second Edition. 'Highly commendable' – *Journal of the Institute of Health Education*

☐ *The Penguin Dictionary of Physical Geography*　£4.95

This book discusses all the main terms used, in over 5,000 entries illustrated with diagrams and meticulously cross-referenced.

☐ *Roget's Thesaurus*　£2.95

Specially adapted for Penguins, Sue Lloyd's acclaimed new version of Roget's original will help you find the right words for your purposes. 'As normal a part of an intelligent household's library as the Bible, Shakespeare or a dictionary' – *Daily Telegraph*

PENGUIN REFERENCE BOOKS

☐ *The New Penguin Dictionary of Music* £4.95

The fourth edition of this comprehensive dictionary covers orchestral, solo, choral and chamber music as well as opera and ballet, and includes detailed entries on composers, instruments of all sorts, orchestras, performers and conductors.

☐ *A Dictionary of Literary Terms* £5.95

Defines over 2,000 literary terms (including lesser known, foreign language and technical terms) explained with illustrations from literature past and present.

☐ *The Penguin Dictionary of Mathematics* £3.95

From algebra to number theory, from statistics to quantum mechanics, this new dictionary takes in all branches of pure and applied mathematics up to first-year university level.

These books should be available at all good bookshops or newsagents, but if you live in the UK or the Republic of Ireland and have difficulty in getting to a bookshop, they can be ordered by post. Please indicate the titles required and fill in the form below.

NAME _____ BLOCK CAPITALS

ADDRESS _____

_____

_____

Enclose a cheque or postal order payable to The Penguin Bookshop to cover the total price of books ordered, plus 50p for postage. Readers in the Republic of Ireland should send £IR equivalent to the sterling prices, plus 67p for postage. Send to: The Penguin Bookshop, 54/56 Bridlesmith Gate, Nottingham, NG1 2GP.

You can also order by phoning (0602) 599295, and quoting your Barclaycard or Access number.

Every effort is made to ensure the accuracy of the price and availability of books at the time of going to press, but it is sometimes necessary to increase prices and in these circumstances retail prices may be shown on the covers of books which may differ from the prices shown in this list or elsewhere. This list is not an offer to supply any book.

This order service is only available to residents in the UK and the Republic of Ireland.